THE TROPICAL AGRICULTURALIST

Series Editor
René Coste
Formerly President of the IRCC

Cocoa

Guy Mossu
Deputy Director of the IRCC – CIRAD

Translated by Shirley Barrett
with editorial advice from
Professor H. D. Tindall
Emeritus Professor of Tropical Agronomy
Cranfield Institute of Technology

MACMILLAN

The opinions expressed in this document and the spellings of proper names and territorial boundaries contained therein are solely the responsibility of the author and in no way involve the official position or the liability of the Agency for Cultural and Technical Cooperation or the Technical Centre for Agricultural and Rural Cooperation.

Original French edition published in 1990 under the title *Le cacaoyer*
in the series *Le Technicien d'Agriculture Tropicale*,
by the ACCT and Maisonneuve et Larose,
15, rue Victor-Cousin, 75005 Paris, France.
No responsibility is taken by the holders of the copyright for any changes to the original French text.

This edition first published 1992
Reprinted 1995

Published by MACMILLAN EDUCATION LTD
London and Basingstoke
Associated companies and representatives in Accra, Banjul, Cairo, Dar es Salaam, Delhi, Freetown, Gaborone, Harare, Hong Kong, Johannesburg, Kampala, Lagos, Lahore, Lusaka, Mexico City, Nairobi, São Paulo, Tokyo

Published in co-operation with the Technical Centre for Agricultural and Rural Co-operation, P.O.B. 380, 6700 AJ Wageningen. The Netherlands.

Printed in Malaysia

ISBN 0–333–57076–6✓

A catalogue record for this book is available from the British Library

Cover illustration: courtesy of the author

The ACP-EEC Technical centre for Agricultural and Rural Cooperation (CTA) operates under the Lomé Convention between Member States of the European Community and the African, Caribbean and Pacific States. CTA collects, disseminates and facilitates the exchange of information on research, training and innovations in the spheres of agriculture and rural development and extension for the benefit of the ACP states. To achieve this, CTA commissions and publishes studies, organises and supports conferences, workshops and seminars, publishes and co-publishes a wide range of books, proceedings, bibliographies and directories, strengthens documentation services in ACP countries, and offers an extensive information service.

Postal address: Postbus 380, 6700 AJ Wageningen, The Netherlands
Telephone: 31 – (0)8380 – 60400
Fax: 31 – (0)8380 – 31052
Telex: 44 – 30169 CTA NL

Agency for Cultural and Technical Cooperation (CTA)

The Agency for Cultural and Technical Cooperation, an intergovernmental organisation set up by the Treaty of Niamey in March 1970, is an association of countries linked by their common usage of the French language, for the purposes of cooperation in the fields of education, culture, science and technology and, more generally, in all matters which contribute to the development of its Member States and to bringing peoples closer together.

The Agency's activities in the fields of scientific and technical cooperation for development are directed primarily towards the preparation, dissemination and exchange of scientific and technical information, drawing up an inventory of and exploiting natural resources, and the socio-economic advancement of young people and rural communities.

Member countries: Belgium, Benin, Burundi, Canada, Central African Republic, Chad, Comoros, Congo, Djibouti, Dominica, France, Gabon, Guinea, Haiti, Côte d'Ivoire, Lebanon, Luxembourg, Mali, Mauritius, Monaco, Niger, Rwanda, Senegal, Seychelles, Togo, Tunisia, Burkina Faso, Vanuatu, Viet Nam, Zaire.

Associated States: Cameroon, Egypt, Guinea Bissau, Laos, Mauritania, Morocco, St Lucia.

Participating governments: New Brunswick, Quebec.

Titles in the Tropical Agriculturalist series

Sheep	ISBN:0-333-52310-5	Ruminant Nutrition	0-333-57073-1
Pigs	0-333-52308-3	Animal Breeding	0-333-57298-X
Goats	0-333-52309-1	Animal Health Vol. 1	0-333-61202-7
Dairying	0-333-52313-X	Animal Health Vol. 2	0-333-57360-9
Poultry	0-333-52306-7	Warm-water Crustaceans	0-333-57462-1
Rabbits	0-333-52311-3	Livestock Production Systems	
Draught Animals	0-333-52307-5		0-333-60012-6

Upland Rice	0-333-44889-8	Sugar Cane	0-333-57075-8
Tea	0-333-54450-1	Maize	0-333-44404-3
Cotton	0-333-47280-2	Plantain Bananas	0-333-44813-8
Weed Control	0-333-54449-8	Coffee Growing	0-333-54451-X
Spice Plants	0-333-57460-5	Food Legumes	0-333-53850-1
Cocoa	0-333-57076-6	Cassava	0-333-47395-7
The Storage of Food		Sorghum	0-333-54452-8
Grains and Seeds	0-333-44827-8		

Other titles published by Macmillan with CTA (co-published in French by Maisonneuve et Larose)

Animal Production in the Tropics and Subtropics	0-333-53818-8
Coffee: The Plant and the Product	0-333-57296-3
The Tropical Vegetable Garden	0-333-57077-4
Controlling Crop Pests and Diseases	0-333-57216-5
Dryland Farming in Africa	0-333-47654-9
The Yam	0-333-57456-7

Land and Life series (co-published with Terres et Vie)

African Gardens and Orchards	0-333-49076-2
Vanishing Land and Water	0-333-44597-X
Ways of Water	0-333-57078-2
Agriculture in African Rural Communities	0-333-44595-3

Contents

Foreword   vi

 1 The cocoa plant and its distribution   1
 2 Species cultivated   7
 3 The plant and its environment   10
 4 Planting material   27
 5 Establishing a cocoa plantation   37
 6 Harvesting and preparation of commercial cocoa   69
 7 Working times per hectare   78
 8 Qualities and defects of commercial cocoa   82
 9 Manufacture of chocolate products   87
10 Nutritional value   92

Glossary   95
Further reading   98
Index   99

Foreword

Apart from J. BRAUDEAU's book *Le cacaoyer*, which first appeared in 1969, the tropical agriculturalist has had no reasonably compact and comprehensive work on the cocoa tree, its cultivation and the preparation and processing of its product – cocoa.

This gap has now been filled thanks to the work carried out by my colleague G. Mossu.

With a wealth of experience acquired over twenty years most of which has been related to the improvement of the cocoa tree, he has been able to make excellent use not only of the results of his own research, but also of those of all the researchers and experts of the Institut de Recherches du Café et du Cacao (IRCC – CIRAD).

No one is better qualified to synthesise all this material. Fully aware of the difficulties encountered in the transfer to the development sector of the knowledge acquired from research, he has succeeded in condensing a wealth of information on the cocoa tree and cocoa while, at the same time, making an operational tool available to the users on the ground.

I am certain that the users of this handbook will be grateful to him for carrying out this task and that cocoa-growers, as well as development chiefs, will gain significant benefit from it.

Marc Belin
Director of the Institut de Recherches du Café et du Cacao

1 The cocoa plant and its distribution

The origin of the cultivated cocoa plant

The cultivated cocoa plant, the botanical name of which is *Theobroma cacoa* L., belongs to the Sterculiaceae family.

It is now generally accepted that the plant originated in several native areas, the most important of these being at the foot of the Andes in the upper reaches of the Amazon river.

The Genus *Theobroma* contains some 22 species, all originating in the tropical rainforests of equatorial America, and some of these species are grown locally for making cooked dishes, jellies or refreshing beverages.

> *However, the only species grown commercially for the production of seeds for chocolate-making or for the extraction of cocoa butter is* Theobroma cacao L.

The history of cocoa-growing

The first evidence of the cultivation of cocoa in Mesoamerica dates back some two thousand years. The Maya Indians were certainly the first to cultivate this tree, the beans of which were used both as a product to be consumed and as currency.

The Aztec Indians of the high Mexican plateaux extended their empire to the cocoa-growing regions where they levied high taxes in the form of seeds which they called '*cacahoatl*' – hence the word 'cocoa'. They attributed divine origin to the cocoa tree, thought to have been brought to earth by the god Quetzacoatl (the plumed serpent). There were many ceremonies in which cocoa was used either as a luxury commodity with sacred value or as a beverage, also sacred, which they called '*xocolatl*', from which the word 'chocolate' was derived.

La Buyeuse de Chocolat

Grayée en manière de Pastel par

Louis-Marin Bonnet

Fig 1 *The chocolate-drinker (eighteenth-century lithograph)* (L. Hayot collection)

It was probably with this legend in mind that Linnaeus gave the cultivated cocoa plant the name *Theobroma cacao*, from the Greek *theos* = gods, and *broma* = food: 'food of the gods'.

When Cortez landed on the coast of Tabasco in 1519 and began the conquest of Mexico, he very quickly became interested in cocoa, more from the point of view of its great value in terms of currency than as a beverage, even though it was always served with great ceremony.

The *xocolatl* of the Aztecs was prepared using roasted seeds which were ground, mixed with water and beaten vigorously. To this mixture was added maize flour and various condiments such as pimento and annatto. When freshly made, this preparation was considered to be nourishing, fortifying and an aphrodisiac, but it was 'pure folly to see how it disgusted those who were not used to it because it had a foam on the surface and froth which resembled wine lees, and those who drank it were brave indeed . . .' (according to Acosta, 1589).

The Spaniards soon discovered that by replacing the pimento and the annatto with cane sugar, which they had introduced into Mexico from the Canary Islands, and with the addition of vanilla, the *xocolatl* clearly became a very acceptable drink which became more and more popular.

Since the end of the sixteenth century cocoa has been grown in most of the tropical regions of Central and South America, as well as in the West Indies. The first cocoa exports to Europe, from Veracruz to Cadiz, date back to 1585. From Spain, cocoa spread through the royal courts of Europe, first to Italy and then to France. Anne of Austria, the daughter of Phillip III of Spain, introduced it following her marriage to Louis XIII in 1615; it was later used in Holland, England and Germany.

Since that period, writers, doctors, planters, tradesmen and gourmets of both sexes have often felt passionately about this 'food of the gods'. Chocolate was a great success in the royal courts of Charles II, Louis XIII in France and Phillip IV in Spain where, it is said, the ladies of the court had chocolate served to them during mass, despite protestations from the bishop. Even theologians found in it a new subject for debate.

The popularity of the beverage led to a great increase in the number of cocoa plantations in the seventeenth century in the New World, particularly Trinidad, Jamaica, Haiti, Venezuela and later in Martinique, where cocoa was originally planted by the French in 1660. It was only introduced into Brazil in 1754 in the Bahia region. The Spaniards, Dutch and Portuguese introduced the crop into all their overseas territories in South East Asia on the one hand, and in the islands of the Gulf of Guinea, off the coast of West Africa, on the other. It was from these islands, Fernando Po (today known as Malabo), São Tomé and Príncipe, that the cocoa plant was introduced to the African continent a little over a century ago.

The production and worldwide marketing of cocoa

Among the major tropical crops, cocoa has never held the dominant role played by sugar cane, tobacco or cotton; it long remained the privilege of the New World, but most of it has been produced in Africa since its introduction there at the end of the nineteenth century.

The twentieth century has seen a considerable increase in production, reaching 1.5 million tonnes in 1964 and exceeding 2 million tonnes today. Africa alone accounts for almost 55 per cent of world production. The large producing countries today are Côte d'Ivoire, Brazil, Ghana, Malaysia, Cameroon and Nigeria. Production has increased at an annual rate of 2 to 2.5 per cent since the middle of the last decade, whereas consumption has increased at a rate of only around 1 per cent.

Each season therefore ends with a surplus, which has the result of permanently keeping prices low. Being a free market subject to the law of supply and demand, the market in cocoa beans has always been characterised by considerable fluctuations from one year to the next, and sometimes during the same year, indeed even during the same month.

Some thirty-seven producing and consuming countries are members of the International Cocoa Organisation (ICCO), the aim of which is to define common rules which should, when applied, enable excessive fluctuations to be limited. To this end, four international agreements have been signed. The fourth, concluded in 1986, uses a range of prices now given in special drawing rights (SDRs), i.e. a basket of currencies in which the dollar is 42 per cent, the Deutschmark 19 per cent, the yen 15 per cent, the French franc 15 per cent, and pound sterling 12 per cent. The use of the SDR aims to minimise the problems caused by the fluctuation of the dollar. Today, the performance of this agreement presents many difficulties which the partners are trying hard to overcome.

Average world consumption of cocoa per person per year is 370 g. It is only 60 g in the developing countries but is 1.25 kg in the industrialised countries. These averages do, however, mask very considerable differences. In the developing countries, annual per capita consumption is 30 g in Africa and 1.60 kg in Colombia. Among the industrialised countries, the world leader is Switzerland with 4.45 kg per person per year, followed by Belgium (3.68 kg), Austria (3.19 kg), France (2.02 kg), the United States (1.99 kg), Sweden (1.71 kg), Australia (1.59 kg), Hungary (1.54 kg), Denmark (1.51 kg) and East Germany (1.45 kg).

The main uses of cocoa

The cocoa bean is the product obtained after the fresh seed has been fermented and dried. It constitutes the raw material of important industries which manufacture:

(1) *semi-finished products* intended for other industries:
- cocoa mass used for making chocolate, biscuits and confectionery;
- melted cocoa, intended for various food industries for sweet products;
- cocoa butter, used in making sweets, chocolate, perfume and in pharmacy;

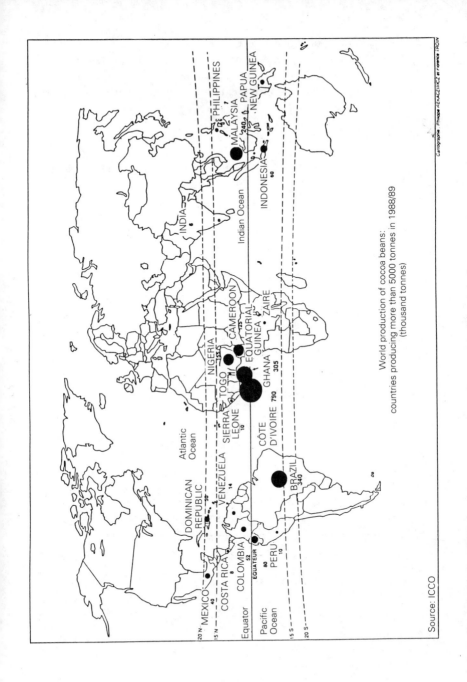

Fig 2 *World production of cocoa beans* (Map: P. Rekacewicz and F. Troin)

(2) *finished products* intended directly for consumption:
• chocolate powder; • bars of chocolate; • chocolate confectionery.
The by-products of this industry – husks, fats extracted from the husks and 'germs' – can be used to feed cattle, manufacture fertilisers, pharmaceutical products and soap.

Table 1 *World production of cocoa (thousand tonnes)*

	1946/47 to 1950/51	1964/65	1974/75	1984/85	1988/89 (Forecast)
Africa	*465.1*	*1 196.3*	*1 015*	*1 080*	*1 431.7*
% of world total, of which:	65.1	79.4	65.3	55.5	59.6
Côte d'Ivoire	45.2	147.5	241	565	790
Cameroon	46.0	91.2	118.0	120	125
Togo	3.5	17.4	15.0	10	14
Ghana	241.4	580.9	381.6	175	305
Nigeria	99.6	298.3	214	151	155
Other countries	29.4	61.0	45.4	59	42.7
Latin America, Caribbean	*245.9*	*285.2*	*479*	*704*	*612.4*
% of world total, of which:	34.4	18.9	30.8	36.2	25.5
Brazil	127.8	118.5	265.5	412	340
Ecuador	21.8	48.2	75.3	120	80
Colombia	9.9	17.5	26	41	52.3
Mexico	7.4	20.6	32	42	40
Dominican Republic	30.3	25.0	33.2	39	50
Other countries	48.7	55.4	47.0	50	50.1
Asia-Oceania	*3.8*	*25.0*	*61*	*161*	*356.7*
% of world total, of which:	0.5	1.7	3.9	8.3	14.9
Malaysia			12.0	93	240
Papua New Guinea	0.2	21.0	36.0	31	36
Indonesia			3.3	22	60
Other countries	3.6	4.0	9.7	15	20.7
World total	**714.8**	**1 506.5**	**1 555**	**1 945**	**2 400.8**

2 Species cultivated (*Theobroma cacao* L.)

Classification of the Theobroma cacao L. *species cultivated*

Like many species found in Amazonia, *Theobroma cacao* L. appears in a great variety of forms. Because of this, for a long time the classification of cultivated cocoa plants was very confused. It was based on the morphological characteristics of the pods, the flowers or the seeds, which all vary enormously.

Today, cocoa can be divided into three large groups, the Criollo, Amazonian Forastero and the Trinitario types. All these types of cocoa plant are interfertile and, by crossing, give fertile hybrids which today represent most of the cultivars used in plantations.

The chromosomal number of *Theobroma cacao* L. is 2n = 20.

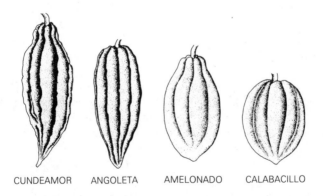

CUNDEAMOR   ANGOLETA   AMELONADO   CALABACILLO

Fig 3   *Main pod shapes*

The Criollo group

Domesticated for a very long time, probably by the Maya Indians, Criollo cocoa trees are now to be found cultivated in Mexico, Nicaragua, Guatemala, Colombia, Venezuela, Madagascar, in the Comoro Islands, Sri Lanka, Indonesia (Java) and on the Samoa islands.

The main characteristics of this group are:
- pale pink staminodes,
- pods which are green or red before ripening, varying in shape but usually resembling the cundeamor type (see Fig. 3),
- generally with a very warty and thin pericarp, and a mesocarp which is only slightly woody, and thin,
- plump beans, almost round in cross-section,
- white or very slightly pigmented cotyledons.

Criollo cocoa is, in fact, very much sought after for its strong aroma and only slight bitterness. It is used in the chocolate industry for luxury products.

Nowadays, Criollo cocoa trees are to be found in isolated groups of trees or in small, very old plantations. They are usually not very vigorous, are slow growing and have small leaves. Furthermore, they are reputed to be very vulnerable to diseases, which is probably one of the reasons why they are being grown less and less and are being replaced by other, more vigorous types.

The original Criollo selection programmes have, unfortunately, not been continued – exceptions are those in Colombia and Venezuela. This type of cocoa is very much appreciated by the consumers, and it should be rehabilitated with new studies being carried out on the existing selections.

The main types of Criollo include Pentagona or Lagarto cocoa, the pods of which are typically pentagonal in cross-section, Real cocoa and Porcelana cocoa.

The Forastero group

This is a very variable group which is found in an indigenous or semi-indigenous state in High Amazonia (Peru, Ecuador and Colombia), in the Amazon basin (Brazil), in the Guyanas and along the Orinoco river in Venezuela. They are now very widely used in plantations throughout all the producing countries.

Their general characteristics are as follows:
- staminodes with purple pigments,
- pods green before ripening, the shape varying enormously,
- thick pericarp and very woody mesocarp,
- more or less flat seeds,
- dark purple cotyledons, yielding a cocoa with a relatively bitter flavour and often acid taste.

The Forasteros constitute almost all of the production currently coming from Brazil, West Africa and South East Asia.

The West African 'Amelonado' cocoa belongs to this group, as do the 'Maranhao', 'Comun' and 'Para' types from Brazil. These last two originated, by mutation, the 'Almeida' and 'Catongo' cocoa types with white cotyledons.

We should also mention the 'Nacional' cocoa type from Ecuador, which produces a fine cocoa with an excellent reputation on the international market under the name of 'Arriba'. Unfortunately, this type of cocoa has almost completely disappeared.

Finally, the Upper Amazonian cocoas, as opposed to the other so-called Lower Amazonian Forasteros, include the Forasteros collected during several prospecting expeditions since the beginning of the twentieth century in the upper part of the Amazon basin, mainly to the west of the town of Iquitos. The Upper Amazonians vary enormously, both in the shape of the pods and in the size and colour of the seeds, which range from dark purple to white. These Upper Amazonians usually bear the name of the place or of the river in the region in which they have traditionally been harvested: Iquitos, Nanay, Parinari, Scavina, Morona, Moquique, etc.

The Trinitario group

This group consists of very different and very heterogeneous types, probably resulting from a Forastero and Criollo cross. The Trinitarios are grown mainly in all the countries where the Criollos were formerly grown (Mexico and Central America, Trinidad, Colombia and Venezuela) as well as in many African and South East Asian countries.

Their botanical characteristics have all the intermediate features of the Criollo and the Forastero groups. They produce a cocoa which is also of intermediate quality.

The Trinitario types which are now found almost everywhere were originally selected in Trinidad, hence their name. The Trinitario cultivars generally bear the name of the bodies or research centres which originally selected them: ICS (selection of the Imperial College in Trinidad), UF (United Fruit selection in Costa Rica), SNC (selection of the Nkoemvone station in Cameroon), etc.

3 The plant and its environment

Morphology and biology

General comments

An adult cocoa tree can grow up to 12 to 15 metres high in the wild state. Its height as well as the leaf area and spread of its branches and leaves depend a great deal on the space available. Consequently, the space which is usually left between the plants when planting allows the adult tree to reach an average height of 5 to 7 metres.

When grown from seed, the cocoa tree is fully developed at about the age of ten years. It is productive, however, well before this age, since the flowers and fruits are formed from the third or fourth year, with full yield generally being produced at around six or seven years. A well-run plantation may continue to be profitable for at least 25 to 30 years.

The root system

The first sign that a cocoa seed has germinated is the appearance, three or four days after sowing, of a whitish root (taproot) which grows rapidly and vertically into the soil. It reaches about 10 to 15 cm in length, 15 to 20 days after sowing. At the same time as this rapid vertical growth phase takes place, lateral roots appear and grow horizontally.

The taproot is fully developed approximately ten years after planting. Its length ranges from 0.8 m to 1.5 or even 2 metres. Along its entire length, lateral roots arise from the taproot but only develop to any great extent in the upper portion, usually between the collar and 15 to 20 centimetres below it.

Up to ten lateral roots may therefore develop in the topsoil, branching frequently to form a mass of root hairs which may cover an area 5 to 6 m in radius around the tree.

> *The root system shows dimorphous growth, characterised by the orthotropic taproot and lateral plagiotropic ramifications, which constitute the lateral roots.*

The aerial system

(a) The trunk

Initially, the stem develops vertically (orthotropically) and the leaves with long petioles appear in accordance with a 3/8 phyllotaxy arrangement. The stem grows in height in successive elongation phases up to the age of eighteen months. Further growth is then interrupted by the degeneration of the terminal bud below which the first branches develop simultaneously, in the form of a whorl of five branches, growing horizontally (plagiotropically). These fan branches form the framework of the tree, and are referred to as the jorquette.

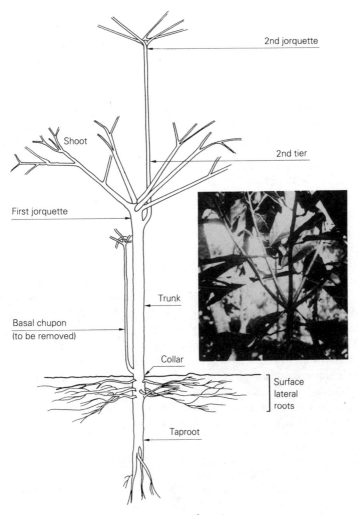

Fig 4   *General development of a cocoa tree in a plantation*

11

At this stage of growth, the trunk will be approximately 1.50 m tall. However, other axillary buds on the trunk, arising from a leaf axil or from a leaf scar immediately below the branches of the jorquette, also develop and produce a large number of orthotropic shoots which behave exactly like the main stem. The most vigorous of the branches is retained after removing the others; this allows the trunk to extend another 1.50 cm in height at which stage a second jorquette will develop. When the latter is well-developed, the first jorquette gradually dies. Several tiers can therefore be successively superimposed on the initial stem. In the wild state, orthotropic suckers will develop freely. Some of them occasionally produce roots, which often gives the wild cocoa plant the appearance of a clump.

In the plantation, the cocoa tree is generally pruned to a single stem at the level of the second jorquette by the systematic removal of all the surplus orthotropic shoots or suckers which may appear on the trunk.

> *The trunk of the cocoa tree, a 3/8 phyllotaxy arrangement, is characterised by: a vertical habit (orthotropy) leaves with long petioles, the development of orthotropic axillary buds, a well-defined growth pattern and the differentiation of normally five plagiotropic buds from below the apex when the terminal bud degenerates.*

(b) The jorquette and the secondary branches

The branches forming the framework of the jorquette, as well as the secondary branches to which they give rise, have a sub-horizontal growth habit. They grow in an undefined and discontinuous manner in successive growth phases called 'flushes'; these are separated by periods when the terminal buds remain dormant. Each flush results in the production of five or six alternate leaves with short petioles. The leaf arrangement (phyllotaxy) is 1/2, i.e. the leaves are alternate on opposite sides of the stem. In general, four to five flushes occur in a year.

All the axillary buds in the axil of each leaf, or leaf scar, may produce a plagiotropic branch. However, orthotropic shoots can also appear on some plagiotropic branches, particularly from the base of those branches which constitute the jorquette.

Approximately seven weeks may elapse between the opening of a bud and the lignification of the wood of the new branch.

> *The branches of a cocoa tree are characterised by: a sub-horizontal (plagiotropic) habit, a 1/2 leaf arrangement, leaves on short petioles, plagiotropic axillary buds (with some exceptions) and an ill-defined growth pattern, with discontinuous flushes.*

(c) The leaf

The young leaves which appear either as the trunk develops or during the flushes, are very often pigmented. Their colour may vary, depending on the tree-type, from pale green, to pink, to deep purple. The young leaves are limp and pendant; as they mature, they become dark green and rigid.

The period of active photosynthesis of the leaf is most marked during the first four to five months of its existence. It then enters a senescent phase, becomes brittle and falls after an average life of one year. The leaves on the various branches of the tree vary in age, since they result from four to five flushes per year.

The leaf stalk, or petiole, which varies in length from 7 to 9 cm for the leaves carried by the orthotropic branches and from 2 to 3 cm for those on the plagiotropic branches, has two characteristic swellings at its base. The lamina, or leaf blade, is entire, simple, oblong, pointed and veined like a feather. The average leaf dimensions are approximately 20 cm long and 8 to 10 cm wide. These may vary considerably and the leaf may be for example, up to 50 cm long depending on the cultivar and the level of exposure to light. The leaves exposed to full light are stronger and thicker than those which are shaded. The very small stomata are only present on the underside of the lamina, the upper epidermis being heavily cutinised.

Flowering

The flowers appear on the bark one or two years after the stem has become lignified. They are not generally seen formed on the trunk before the development of the jorquette. The first flowering may occur at the age of two years for very precocious trees, but it usually appears three to four years after germination.

A succession of flowering occurs, depending on the environmental conditions and the physiological condition of the plant. With few exceptions, the cocoa plants of one plantation, of one country or of an entire region flower at the same time, in successive phases; the intensity of flowering is variable and depends on condition of the trees. Some cocoa trees may have pronounced flowering peaks, in general twice a year, separated by small but continuous flowering periods, as in the Upper Amazonian selections. Others, such as the West African Amelonados, have maximum flowering peaks which are separated by periods when no flowers are produced.

The flowers are grouped in inflorescences which develop from the axillary buds after the leaves have fallen and as soon as favourable physiological, soil and climatic conditions exist. Each inflorescence is a dichasial cyme with very short branches 1 to 2 mm long.

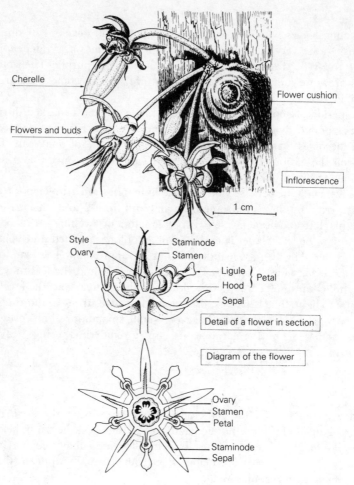

Fig 5 *Inflorescence and flower of the cocoa tree* (After N. Hallé)

The axillary buds which have developed into inflorescences retain this function permanently. Their development each year, at the same sites on the bark, ultimately produces pronounced swellings which are called 'flower cushions'. A flower cushion can carry numerous flowers at any one time. Whatever the period of flower initiation, the flowers are produced simultaneously on all the flower cushions of the tree.

Each flower is supported by a pedicel 1 to 3 cm long. The flowers are hermaphrodite, small (between 0.5 and 1 cm in diameter), regular and pentamerous, i.e. the flower parts are in fives. The five sepals, which are joined at the base, are white or tinged with pink. The five petals, which alternate with the sepals, have a very characteristic shape. They are very

Fig 6 *Left   Flower buds, flower and first stage in the formation of the fruit* (Photo: G. Blaha)

Fig 7 *Right   Flowering and fruiting* (Photo: G. Blaha)

narrow at the base; they then widen and become concave, forming a small white hood edged internally with two violet veins; the hood is open towards the axis of the flower. The upper part of the hood is extended by a thin, tongue-like structure which recurves towards the exterior, and the end widens into a lanceolate ligule which varies in colour from white to bright yellow, depending on the type of tree.

The superior ovary consists of five loculi, each containing six to twelve ovules arranged around the central axis of the ovary. The tubular branched style ends in five stigmata.

The androecium consists of five stamens alternating with five sterile staminodes. Stamens and staminodes are joined at the base, forming a short tubular sheath. The purple-coloured staminodes stand upright around the style, whereas the white stamens are recurved towards the outside; the anthers are therefore enclosed within the hoods of the petals. Each stamen is divided into two and the anthers each have four pollen sacs.

As soon as the flower opens, the dehiscence of the anthers releases the pollen, which is immediately functional, but, under normal conditions, only remains viable for a maximum of 48 hours. The flower bud begins to open in the afternoon and flowering is complete by the early hours of the following morning. A flower bud about to open is recognisable by its rounded shape and by the marked lines separating the sepals.

Fruiting

(a) Pollination

Such a complicated flower is not easy to pollinate. Pollination is mainly by insects, but the insects responsible, since they are very small, are

15

extremely difficult to detect in the field. In most instances, the main pollinating agents, identified by trapping, are midges of the genus *Forcypomyia* of the *Ceratopogonidae* family. Other pollinating agents are ants of the genus *Crematogaster*, the diptera cecidomyiidae, thrips and leaf-hoppers.

Almost 60 per cent of the flowers produced by the cocoa plant are not pollinated and drop after about 48 hours. Only about 5 per cent of the flowers pollinated receive adequate quantities of pollen grains to fertilise all the ovules. This chronic under-pollination, which has been confirmed and measured in several countries in Africa and America, is partly due to environmental factors but is particularly influenced by the number and activity of the pollinating insects. It is also due to the quantity of pollen produced by the stamens, which varies with time and the health of the tree. An adult cocoa tree produces thousands of flowers each year, but only a small percentage of these reach the first stage of fruit-formation.

(b) Fertilisation and incompatibility
Successful pollination is considered to have taken place when, after three days, the flower is in 'swollen ovary' state, which is the first visible sign that the ovules have been fertilised.

Incompatibility may, however, become apparent up to several weeks after pollination, therefore leading to dropping of the young fruit. Incompatibility of the type which occurs in cocoa has characteristics rarely encountered in other plant species. In most plant species, the incompatibility occurs between the pollen and the style of the flower due to the inability of the pollen to germinate and enter the tissue of the style. In cocoa, the pollen is always capable of germinating; the incompatibility reaction only occurs later, in both the ovary and ovule.

It has been suggested that these phenomena are produced genetically. Their intensity varies according to the origins of the tree. Understanding this mechanism, as for any cultivated species, is of vital importance to any selection programme carried out on the plant.

The Upper Amazonian Forastero cultivars are generally self-incompatible but are almost always intercompatible. The Trinitarios include a large proportion of self-incompatible cultivars which, unlike the Upper Amazonians, only accept pollen from self-compatible trees to achieve satisfactory fertilisation. The Lower Amazonian Forastero cultivars and, in particular, the Amelonados, are normally self-compatible.

The fruit

Taking into account the complicated arrangement of the parts of the flower, the normal level of under-pollination and the compatibility problems, the level of successful fertilisation may vary considerably. When it has been completed, however, the fruit will begin to develop from the third day after pollination, i.e. at the 'swollen ovary' stage.

The fruit of cocoa is normally referred to as the 'cherelle' during its early growth; it becomes a 'pod' when it has reached its final size, and is considered to be mature after five to six months, depending on its origins. The pod contains a single cavity in which the seeds, surrounded by a thick mucilaginous pulp, appear to overlap one another in five longitudinal rows.

The fruit is attached to the branch or trunk by a woody stalk which was originally the stalk of the flower but which thickens as the pod develops. The pericarp or cortex of the pod consists of three distinct layers; the hairy and thick epicarp, which is more or less hard (the epidermis of which may be pigmented); the mesocarp, which is thin and hard and more or less woody; and the hairy endocarp, which is of varying thickness.

Pods generally contain an average of thirty to forty seeds, an estimate which is convenient for a rough evaluation of yield in the plantation. However, the number of seeds per pod varies enormously. This variation depends on factors recently quantified by the IRCC, namely:
- the level of pollination, as shown by the effective deposition of pollen on the styles of the flowers;

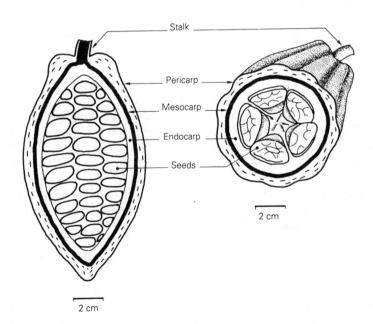

Stalk

Pericarp

Mesocarp

Endocarp

Seeds

2 cm

2 cm

Fig 8 *Longitudinal and transverse sections of a cocoa pod*

17

- the average number of ovules in the ovaries. This is a very important clonal characteristic as it represents the plant's potential to produce seeds;
- fertility, i.e. the numbers of fertilised ovules which develop into seeds;
- finally, the minimum number of seeds necessary for a cherelle to remain actively growing. In actual fact, the probability of the young cherelle remaining on the tree decreases very rapidly if the quantity of seeds which it contains is lower than a threshold number. This is called the 'differential wilting point', which varies according to the origin of the cultivar.

(a) Wilting of the young fruit

Irrespective of disease or insect attack, young fruit may be lost due to the cherelle wilting on the tree. This phenomenon ('wilt'), generally appears between fifty and seventy days after pollination, and can affect 20 per cent to 90 per cent of the cherelles formed. Several reasons for this have been given, among which are:

- that it is the manifestation of a physiological mechanism which regulates fruiting, and which is probably controlled by growth hormones. This is referred to as 'physiological wilt';
- incompatibility phenomena which can become apparent at a later stage;
- differential wilting. The stage of development of the fruit at which this occurs is not known.

The critical period in question occurs, together with other chemical or physiological changes in the fruit, at the same time as the initial cell divisions of the embryos in the seeds occur.

(b) Shape, size and colour of pods

These characteristics vary widely, depending on the genotypes involved.

Before it ripens, the pod may be either green, more or less dark, reddish purple or green with a reddish-purple tinge. As a general rule, all the Forastero pods are green before they ripen and yellow when they are fully mature. A reddish-purple pod pigmentation is a characteristic of the Criollo types, and this is becoming more or less marked in the Trinitario types, which are hybrids between the Criollos and Forasteros. This initial reddish-purple pigmentation becomes bright red or orange when the pod is mature.

The shape of the pod is determined by the ratio between the pod length and width, and also by the shape of the two ends of the pod. Pod shape can vary enormously, from the almost spherical shape of the 'calabacillo' types to the elongated and pointed shape of the 'cundeamors' or of the 'angolettas', with the oval and regular shape of the 'amelonados' as an intermediate form. The stalk end may be restricted in the form of a bottleneck, while the distal end may be more or less pointed.

Five to ten regularly-spaced furrows generally run along the length of the pod from one end to the other. The surface of the fruit may be either completely smooth or extremely warty.

At maturity, depending on the cultivar, the pod length varies between 10 and 35 cm. On the average, it would vary between 15 and 30 cm in length. The weight is also extremely variable, and may range from 200 g to more than 1 kg, but an average pod weight would be in the region of 400 or 500 g. An average pod contains 100 to 120 g of seeds. All of these average measurements conceal very significant variations.

The seed or 'fresh bean'

(a) Morphology

The cocoa seed or fresh bean is shaped rather like a plump almond, and is surrounded by a white mucilaginous pulp which is both sweet and rather sour. The average dimensions of the seed are 20 to 30 mm in length, 12 to 16 mm wide and 7 to 12 mm thick.

The seed has no endosperm and, from the outside to the centre, consists of:

- a thin, resistant, pink, veined husk, originating from the tissues comprising the seed coats of the ovule;
- a fine skin, the silver skin, which is translucent and shiny;
- the two cotyledons which, inside this skin, occupy the total volume of the seed. The cotyledons vary in colour, according to their origins, from white to dark purple, and are tightly folded with many lobes overlapping one over the other. Between the two cotyledons, which are joined at their base by a 6 to 7 mm radicle (primary root), is a rudimentary plumule (shoot). The radicle and plumule together form the embryo, and are often incorrectly called the 'germ' of the cocoa bean.

The size and shape of the seeds and the colour of the cotyledons are distinctive characteristics of the individual cultivars. Although they are mainly controlled by genetic factors, these characteristics may often be accentuated by climatic factors.

After the pulp and the outer husk have been removed, the seed generally weighs between 1.3 g and 2.3 g. The same seed, after drying, will weigh between 0.9 and 1.5 g. The average weight of the husk, compared with the total weight of the depulped and dried seed, is generally between 5 and 8 per cent of the total weight. The average weight of the pulp, compared with the total weight of the seed, is generally less than 40 per cent of the total weight. Finally, the weight of the dried cotyledons is generally 65 per cent of the weight of the fresh cotyledons, but this figure may vary from 50 per cent to 85 per cent.

19

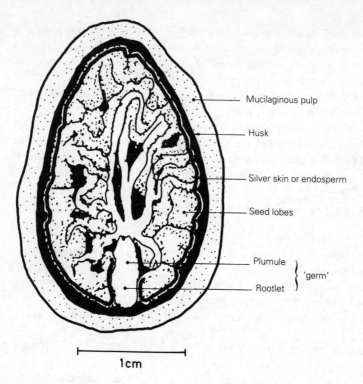

Labels on figure:
- Mucilaginous pulp
- Husk
- Silver skin or endosperm
- Seed lobes
- Plumule } 'germ'
- Rootlet

1cm

Fig 9 *Longitudinal section of a cocoa seed*

(b) Anatomical structure and chemical composition of the cotyledons

A histological study of the fresh cotyledons reveals three types of cell:

- epidermal cells, which are arranged in a monocellular layer;
- reserve parenchymatous cells, which constitute approximately 90 per cent of the total cotyledon tissue. These cells, which are colourless, contain fatty compounds, referred to as 'cocoa butter', and proteins, in the form of granules of aleurone, and starch. The cotyledons are very rich in fatty material, which in general constitutes between 45 per cent and 65 per cent of the total weight of the dried seeds, depending on their genetic origins;
- pigment-containing cells which make up approximately 10 per cent of the tissues of the cotyledons and give them their distinctive colour. These cells contain many polyphenols (tannins, cathechin, anthocyanin, leucoanthocyanin) and purines (theobromine and caffeine). The anthocyanin content, commonly called 'cocoa red' or 'cocoa purple', gives the cotyledons their purple colour. The tannins, which have an astringent taste, oxidise rapidly, giving coloured products which are commonly referred to as 'cocoa brown'.

Theobromine (3–7-dimethylxanthine) and caffeine (1–3–7-trime-thylxanthine), which are more or less bonded to the tannins, form complex compounds in the fresh cotyledons. Theobromine, among other compounds, is responsible for the bitterness of the cocoa beans, i.e. of the seeds which have undergone the necessary fermentation and drying in the preparation of commercial cocoa.

> *No substance which gives the characteristic 'chocolate flavour' has been found in the unfermented cocoa bean.*

(c) Germination of the seed and germination potential

The germination of the cocoa seed is epigeal. Four to six days after sowing, the root rapidly extends and grows vertically into the soil while a vigorous, rapidly elongating hypocotyl carries the cotyledons upwards to the soil surface. Initially arching downwards while the cotyledons remain below the soil surface, the hypocotyl gradually becomes erect and raises the cotyledons to a height of 8 to 12 cm at the end of 10 to 12

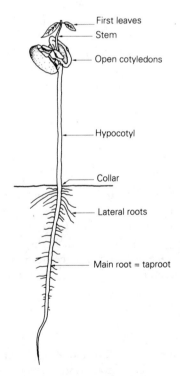

Fig 10   *Germination of the cocoa seed (10 to 12 days after sowing)*

days. At this stage, the two cotyledons separate and bend at a pronounced angle from the hypocotyl axis. The plumule develops into a stem on which, approximately 15 days after sowing, the first four real leaves appear. The very short internodes between the leaves give the appearance of a rosette formation. The terminal bud continues to develop and the stem grows vertically, eventually becoming the trunk of the young cocoa tree.

The seed becomes physiologically mature long before the fruit ripens and, under normal harvesting conditions, the seed should be sown as soon as it has been removed from the pod. The germination potential of the seed then decreases rapidly, particularly if the relative humidity of the atmosphere in which the seed is stored is rather low. A simple and inexpensive method of retaining the germination capacity of seeds intended for local distribution consists of treating each pod, under conditions of low humidity, with a mixture of 90 per cent talc and 10 per cent benomyl (Benlate). When the freshly-harvested pods are thoroughly immersed in this mixture, owing to the surface humidity of the pericarp, the pods will be completely covered. The cortex will continue to lose surface moisture, while the humidity within the pod remains at an appropriately high level. This procedure ensures that 90 per cent of the germination potential of the seeds will be retained for approximately one month. The pods can then be divided into lots of 50 or 60 pods, which is the quantity needed to plant one hectare. These are then transferred to packing cases which can be easily stacked on a covered vehicle. Transporting the fruits in this way prevents their being damaged by bruising or compression.

Ecology

Many ecological factors are involved if the cultivated cocoa tree is to grow at an optimum rate, bear a large number of flowers and fruits and produce normal and well-distributed flushes throughout the year. Due to the complexity of the manner in which they interact, it is difficult to dissociate the influence of individual ecological elements from that of all the other environmental factors, even with the application of modern technology.

Climatic factors

The climate, which is assumed to include all meteorological phenomena which influence the ambient conditions, has a direct effect on the morphology, growth, fruiting and general well-being of the cocoa plant.

(a) Temperature

Cocoa plants respond well to a relatively high temperature with a maximum annual average of 30–32°C and a minimum average of 18–21°C. The monthly average of the minimum daytime temperature should exceed 15°C and the absolute minimum is 10°C; below this temperature, trees are likely to suffer severe damage.

(b) Rainfall

Variations in the yield of cocoa trees from year to year are affected more by rainfall than by any other climatic factor. Trees are very sensitive to a soil-water deficiency, particularly when they are in competition with other plants, shade trees or casual weeds, a situation which frequently occurs in plantations.

Rainfall should be plentiful, but it is particularly important that it is well distributed throughout the year. An annual rainfall level of between 1500 and 2000 mm is generally preferred, provided that if, during the dry season, there is less than 100 mm of rain per month, this season does not exceed three months.

There are special situations in which, when the rainfall is considerably lower than 100 mm per month, irrigation can be used to advantage.

(c) Atmospheric humidity

A hot and humid atmosphere is essential for the optimum development of cocoa trees. Relative humidity is generally high in the cocoa-producing regions. It may often be as high as 100 per cent at night, falling to 70–80 per cent during the day, and sometimes lower in the dry season, during which periods the leaves become limp and droop. It is important to limit the effect of the drying winds, such as the Harmattan in West Africa, by using wind-breaks, shade trees or even by adopting high-density planting.

(d) Light and shade

The cocoa plant originated in the Amazonian forest. Consequently, it has traditionally been grown under shade, such as the natural shade provided by African forest trees, the artificial shading provided by fast-growing shade trees in America, the species of which vary from country to country, with a marked preference for leguminous plants, and, finally, the artificial shading provided by the coconut palm in South East Asia and in the South Pacific.

Owing to its photosynthetic potential, which allows the tree to make optimum use of any light available, the cocoa tree cannot be considered to be a typical shade-loving plant. However, it will grow well under very dense shade which, provided that the temperature remains close to the optimum (32°C), will have no effect on its photosynthetic potential. This quality of its being so well-adapted to shaded conditions means that it cannot also be considered as a light-loving plant. The ambiguous situation regarding the need of the cocoa plant for shade has resulted in many arguments concerning the need for shade and the nature of the shading to be provided.

It is now accepted that shading is a factor which can limit production, but only – and this restriction is very important – when all the other environmental factors are favourable, i.e. optimum temperatures, excellent average humidity, an optimum level of rainfall, a humus-rich soil enriched with fertiliser and well-established insecticidal and fungicidal treatments.

The decision to keep or remove shading in the adult cocoa plantation can only be made by the planter himself who, depending on his means, the size of the plantation and the time he has available, should be able to determine his ability to control all the factors which would promote the desired development of his trees.

Retaining permanent shading reduces some of the risks to which unshaded trees are liable and ensures a regular level of production, even if this is less than the yield which could be achieved without shading. Permanent shading should be adjusted progressively to allow the penetration of a maximum level of 50 per cent of the total light available. It may be reduced even more allowing access to 75 per cent of the total light, if the cocoa trees themselves form a dense and regular canopy, if the soil is fertile and has an adequate supply of water, and if the rainfall is regularly distributed throughout the year.

> *Temporary shading is indispensable during the cocoa tree's early years. The temporary shading should be relatively dense, allowing through no more than 50 per cent of the total light for at least two years after planting. It should be progressively reduced as the cocoa tree develops, but never before the jorquettes have been properly formed.*

The soil

The soil is important because of its physical and chemical characteristics, which are closely linked with the climatic conditions. The cocoa plant can grow in a wide range of soils but, given equivalent climatic conditions, the deeper and richer soils are far more favourable to the development and production of the trees.

(a) Physical properties
- The soil must be at least 1.5 m deep. This is particularly important if there is insufficient or poorly distributed rainfall.

- The soil structure must be as homogeneous as possible, to allow the roots to penetrate easily. Stones or gravel do not constitute a serious problem unless they are present in very large quantities.
- The texture of the soil in which the trees are planted has to satisfy two sometimes contradictory requirements. It must, on the one hand, have good water-retaining properties but must also be well-aerated and have good drainage. The cocoa tree is particularly sensitive to a lack of water but is equally sensitive to soils which are inadequately aerated due to poor drainage.

The ideal soil texture is closely linked with the other ecological factors and, in particular, with rainfall:

- in low rainfall areas sandy soils, which are too permeable, are not suitable.
- in high rainfall areas a sandy soil can be selected, provided that the area is not subject to prolonged hot, dry, seasons. A very sandy soil may be acceptable, particularly if it contains an adequate supply of organic matter, but a soil with a sandy clay texture is always preferable.

(b) Chemical properties

It is thought that the chemical properties of the topsoil are the most important, taking into account both the large number of lateral roots which grow at this level, and the basic function of these creeping roots, which is generally to absorb minerals.

(c) pH (level of acidity)

Cocoa plants can grow in soils with a very variable pH which may range from very acid (pH 5) to very alkaline (pH 8). However, most good cocoa-growing soils are close to neutral (pH 7), with the optimum pH being slightly acid (pH 6.5).

In the more acid soils the main nutrient elements, particularly phosphorus (P), become less usable by the plant, whereas the trace elements, such as iron (Fe), manganese (Mn), copper (Cu) and zinc (Zn), may sometimes reach toxic levels. Although they are relatively rare in tropical environments, alkaline soils (where the pH is higher than pH 7) are often deficient in trace elements; deficiencies such as zinc deficiency can have very serious effects on the growth of the cocoa plant.

(d) Organic matter content

A high content of organic matter in the topsoil is essential for good growth and good productivity. A content of 3.5 per cent in the top 15 centimetres of the soil should be regarded as being the minimum amount required. The organic material is generally concentrated in the topsoil, which should be protected from erosion by heavy mechanical machinery and rainfall and also from degradation by prolonged direct sunlight during the preparation of the land.

(e) Nutrient element content

Cocoa-growing soils must have certain anionic and cationic balances which, as far as is known at present, are as follows:

- the optimum total nitrogen/total phosphorus ratio should be close to 1.5, with the assimilable phosphorus content being at least equal to 180 ppm of P or 0.229 per thousand of P_2O_5;
- the exchangeable bases are balanced at 8 per cent potassium (K), 68 per cent calcium (Ca) and 24 per cent magnesium (Mg);
- the optimum balance of these exchangeable bases with the total nitrogen content must follow a linear regression according to the formula $S = 8.9\,N - 6.15$;
- the minimum saturation level of the exchangeable bases must be more than 60 per cent.

It must be stressed that all this data relates to the topsoil in which the fibrous roots develop. However, further studies on the absorption capacities of the different parts of the root system would certainly lead to a better understanding of the often disappointing response of the cocoa plant to fertiliser applied to the surface. A number of practical observations, such as those relating to the particularly beneficial action of high rainfall (or of irrigation) in making mineral fertilisers more available, indicate that the absorption functions of the deeper roots, such as the taproot and lower branching roots, may play a not inconsiderable role in the mineral nutrition of the cocoa plant. This emphasises the value of embarking upon or continuing research into the physiology of the cocoa plant.

4 Planting material

Selecting the cocoa plant

Most of the cocoa-producing countries have agricultural research centres in which specialists evaluate and suggest the most appropriate types of plant for local climatic conditions and local growing methods.

Very roughly, the criteria used to select the cocoa plant are vigour, precociousness, productivity, the size and quality of the beans and behaviour when subjected to disease and insect attack.

This selection is carried out in the following progressive stages: the establishment of a collection, the selection of clones from this collection, the hybridisation and evaluation of the interclonal descendants and, finally, the distribution of the best of these descendants.

The collection

The cocoa tree collection, which is a veritable gene bank for the selector, contains not only trees logged in the National Cocoa Collection and in local or foreign research stations, but also the largest possible number of trees introduced from different regions where the plant originates. It therefore contains Forasteros from Amazonia (Upper and Lower Amazonians), Criollos from Central America and Trinitarios from the northern regions of South America. As well as these *Theobroma cacao* L. species, there are also other *Theobroma* species, which may be the subject of special genetic studies.

Clonal selection

The second stage of the programme is the evaluation of the trees in the collection. The individuals are propagated vegetatively and are therefore made up of different clones, which are then established in clonal trials to compare their agricultural behaviour in terms of yield and vegetative vigour.

At the same time, their genetic characteristics are studied, concentrating in particular on the rhythm and intensity of flowering, the size and quality of the beans and their response to disease and insect attack. Recently, two new genetic characteristics were defined: the average number of ovules per ovary, which represents the potential production of seeds per pod, and the differential wilting point, which represents the minimum number of seeds necessary for the fruit to remain attached to, and to develop on, the tree. Finally, in the laboratory, research using markers, carried out by analysing the enzymatic systems in electrophoresis, means that the genetic variability of material from different origins can be studied.

This clonal selection procedure introduces the possibility of using listed and catalogued cultivars whose qualities it may be possible to combine by hybridisation.

In some countries where the collections and populations of cocoa trees are very heterogeneous, with the Trinitario type predominating, clonal selection soon reveals which trees perform best. They are multiplied by vegetative means, i.e. by cuttings, but particularly by grafting, and are used directly in the plantation.

Multiplication by vegetative propagation

In general, the cocoa plant is well-suited to vegetative propagation by horticultural cuttings or by grafting. On the other hand, micropropagation techniques involving *in vitro* cultivation have not yet given particularly convincing results. In this area, the cocoa plant is proving to be difficult; a great deal more research is needed. The *in vitro* techniques would in fact offer high multiplication possibilities, under sterile growing conditions. It should then be possible to create micro-collections and gene banks which take up little space, thus facilitating exchanges between countries.

Propagation from cuttings of plagiotropic branches is a technique used mainly for planting plots for the collection and clone plots at the research stations. Plants from plagiotropic cuttings are no longer distributed to the planters as planting material. A new technique of taking cuttings from the orthotropic branches is currently being studied by the IRCC and should soon be operational.

Vegetative propagation by grafting, which does not require large investment, is, on the other hand, favoured in several countries in America and in South East Asia, owing to the presence of a skilled and fast-working workforce, which is a factor in favour of this method of propagation.

Hybridisation and selecting hybrids

The creation and evaluation of interclonal hybrids is the next stage. This involves selection on a generative basis. As the cocoa plant is highly

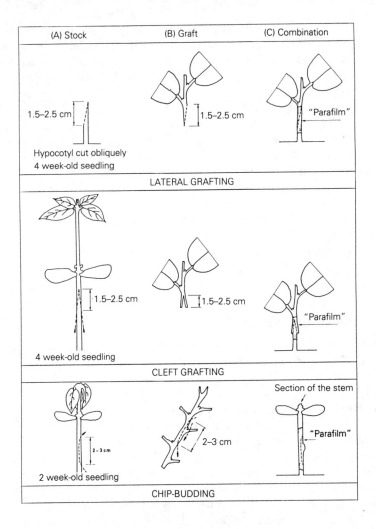

(A) Stock	(B) Graft	(C) Combination

1.5–2.5 cm

Hypocotyl cut obliquely
4 week-old seedling

1.5–2.5 cm

"Parafilm"

LATERAL GRAFTING

4 week-old seedling

1.5–2.5 cm

1.5–2.5 cm

"Parafilm"

CLEFT GRAFTING

2 week-old seedling

2–3 cm

2–3 cm

Section of the stem

"Parafilm"

CHIP-BUDDING

Fig 11   *Some grafting techniques for establishing clonal plots*

heterozygotic, the search for the best possible combination of characteristics requires a large number of crosses to be made between selected
clones.

> *The most spectacular results were obtained from Upper Amazonian clones*
> *which, when crossed with other plants of Forastero or Trinitario origin,*
> *revealed exceptional hybrid vigour from the first generation onwards.*

The seedlings resulting from these crosses produce very precocious plants with yields which are comparable to and often higher than those of the best clones used hitherto. This particular characteristic has been exploited in all the programmes for improving the propagating material of the producing countries.

The families obtained have a production potential of between 1 and 3 tonnes of commercial cocoa per hectare from the fourth year after planting, whereas the average yield of adult plantations originating from non-selected material varies between 250 and 500 kg per hectare.

These hybrid selections are currently being distributed on a large scale to create new plantations. The hybridisation and hybrid selection programme is continuing year after year with new clones being added to the collections. The widening of the genetic range available is a permanent priority of the selector who wishes to exploit the potential of this plant even further by bringing together the best characteristics through hybridisation.

Distribution of the best hybrid selections

Crosses between the parent clones to obtain selected hybrids are made on a commercial scale in specially prepared fields called 'seed gardens'.

(a) Seed gardens with natural pollination

The hybridisation policy of these seed gardens is based on the self-incompatibility of the selected clone used as the female parent, usually an Upper Amazonian, and consists of placing the cuttings of the two parent clones in the same plot. The pods produced, which will have resulted from uncontrolled pollination, are then taken from the female plant, the seeds of which, because of the self-incompatibility of the tree, can in theory only arise from one effective cross-pollination, normally between the two adjacent trees.

The planting and maintenance of a seed garden have the same agricultural and health requirements as those of a conventional cocoa plantation.

(b) Seed gardens with hand-pollination

Hybrid seed production is nowadays most frequently carried out by controlled pollinations, by hand, between two parent clones which are planted in separate plots. This gives several female clones and, quite separately, the male clones. An arrangement of this kind offers numerous possibilities for controlled crosses and, additionally, great flexibility in the choice of the combinations which can be made.

The inter-state transfer of vegetative material

Because of the location of plants infected with particularly serious diseases in geographical zones which are clearly identified, such as witches' broom (*Crinipellis perniciosa*) on the American continent, or 'swollen shoot' (virus) in Africa, precautions must be taken during any transfer of plant material from one country to another, from one continent to another or even from one region to another.

These transfers must always be supervised by or carried out with the collaboration of the research stations or of authorised organisations.

The official recommendations for cocoa, published by the International Board for Plant Genetic Resources (IBPGR) are:

1 Shipping pods or rooted plants is not recommended.

2 Material should be moved as seeds whenever possible, according to the following procedure:

(a) Collect apparently healthy seed pods.

(b) Open pods and discard all immediately suspect pod and seed material.

(c) Remove pulp from seeds and reject all seeds from a pod if any seeds appear defective.

(d) Remove surface moisture from the seeds and treat them with appropriate fungicides and insecticides according to manufacturers' instructions.

(e) Pack in a suitable medium, such as sterile sawdust. Outer packing material should be sterilised and treated with a contact insecticide. Packing should be carried out in a clean, covered area out of the wind.

(f) Seeds should be subjected to strict post-entry quarantine following arrival in the recipient country for the duration of at least three growth flushes.

A simple and economic method has been devised by the IRCC which consists of depulping the seeds using enzymes. This process can be carried out on large quantities of fresh seeds using an agitator with a rotating drum. The seeds are rinsed and drained and are then rotated in a powdery mixture of slaked lime, charcoal and fungicides. They are then lightly dusted and packed in batches of 850 to 1000 seeds in isothermic and watertight packaging. This method helps to retain the germination potential of 85 per cent of the seeds packaged in this way for from 10 to 15 days of transportation and storage.

Vegetative material, in any form, cannot be transported directly to the country of destination. It must, by law, remain in intermediary quarantine in a non-producing country.

31

3   The procedure for transfer of vegetative material is as follows:

(a) Budwood should be selected from healthy plants and washed in soapy water. It should then be dipped in a water solution or suspension of a combination of an appropriate fungicide and insecticide according to the manufacturers' instructions.

(b) The ends of the budsticks should then be dipped in paraffin wax to prevent dessication.

(c) Each budwood stick should be labelled clearly so that no confusion may occur, and each accession should be packed separately.

(d) Budsticks should be incinerated in intermediate quarantine after removal of the buds.

(e) Following arrival in the recipient country the budded material should be kept under close observation for at least three growth flushes.

(f) Material originating in countries where virus diseases are known to occur should be indexed during intermediate quarantine.

(g) Material sent from intermediate quarantine to the recipient country should undergo post-entry quarantine for at least three growth flushes.

It is recommended that the leaves are not removed from the grafting sticks. In fact, the axillary buds of these leaves can be preserved in excellent condition for grafting if the petiole and two-thirds of the lamina are retained. This is then treated by spraying with an anti-transpirant solution such as 5 per cent Folicote. Furthermore, this material will keep better during transportation if care is taken to wrap each stick with 'parafilm' (a thin, transparent plastic film) or, failing this, with damp newspaper.

4   Unpacking of seeds and budwood should be carried out in an enclosed area to contain any escaping insects. Appropriate protective clothing should be worn to prevent skin/eye contact and breathing of chemicals from the packing material. The packing material should be destroyed.

Production of the seedlings and plants: the nursery

The nursery is the place where the seeds are germinated as soon as they are received and where the young plants are raised for five to seven months, sometimes more, with a view to their being planted out in the field. For the planter, it is a very important place, upon which the success or failure of the future plantation will to a large extent depend.

There are many points in favour of sowing the seeds in the nursery, as opposed to directly out in the field, as shown in Table 2.

Table 2 *Sowing the seed in the nursery or out in the field*

Advantages of the nursery	Disadvantages of direct sowing
It saves time. While the plants are being raised in the nursery, the ground can be prepared in the plantation.	Very high loss of seeds.
Protection of and monitoring the health of the young plants.	Many uncontrollable attacks (caterpillars, rodents, mammals) and frequent destruction of the young plants when the plot is being weeded.
Watering guaranteed.	Water requirements subject to the vagaries of the climate.
Plants can be selected at the time of planting so as to be certain that the seedlings will 'take' in the plantation.	No selection of the plants and far slower development of the plantation.
The best planting period can be selected.	

The best site for the nursery

The nursery must be sited on flat or very gently sloping ground which is well drained and not susceptible to flooding, close to a permanent water source and as close as possible to a track suitable for motor vehicles, and as near as possible to the plantation. It must be checked that there are sufficient quantities of good earth available close by. Finally, in regions which are susceptible to high winds (tornados, and so on) care must be taken not to place the nursery too close to the edge of the forest.

Land clearance

The site chosen must be cleared of any vegetation, the ground must be thoroughly weeded and, if necessary, drainage channels must be dug in the direction of the steepest slope. A framework of wooden or bamboo poles 3 m high should be built: this will support shading, at a distance of some 2.5 m from the ground, letting through approximately 50 per cent of the total light. Shading of this kind can easily be created using palm fronds. These dry slowly, and progressively allow in the external light; this will acclimatise and harden-off the young seedlings before they are planted out in the field. In several countries, shading of the nursery is achieved using various permanent plants such as *Hevea*, the oil palm or *Gliricidia*. Lateral protection is often indispensable to supplement the shading if the site is very exposed and to avoid any interference by animals.

An area of 80 m² of nursery will be needed to produce sufficient seeds to plant one hectare.

Preparation of containers for sowing

The containers most often used are polythene bags or sleeves, perforated in their lower half and 25 cm high. They are cut from a black plastic sleeve 12 cm in diameter and 35 μ thick. The bags are filled with sifted soil containing humus, which may be enriched with clay (20 to 30 per cent) if the soil is too sandy, or mixed with river-sand if the soil is too heavy. Care must be taken throughout the growth stages of the young seedling always to maintain the level of the soil in the bag, so that the plastic does not curl over inwards, thereby rendering watering ineffective.

Arrangement of the beds

The bags are arranged in beds 1.20 m wide, made up of ten bags at the front, in series of double rows separated from each other by a space of 20 cm. Arranged in this way, the beds are separated by paths 60 cm wide allowing easy access for cultural operations or wheelbarrow access.

Seed preparation

The seed must be sown as quickly as possible and, at the latest, two days after the pods have been harvested, because their viability decreases rapidly. After they have been removed from the pod, the seeds are depulped by washing them in plenty of water then rubbing them in fine sand. This process, which may appear time-consuming and fastidious, avoids the possibility of contamination of the sweet and slightly acidic substrate of the pulp. The seeds which are going to be sown must be of normal aspect. Flat seeds, excessively small seeds and germinated seeds are to be discarded.

Sowing

Each seed is sown in a bag in soil which will have been copiously watered the previous day. The seed is placed flat on its side in the centre of the soil in the bag and then pushed down to a depth of one centimetre and covered with soil. The bag is watered again immediately after sowing.

34

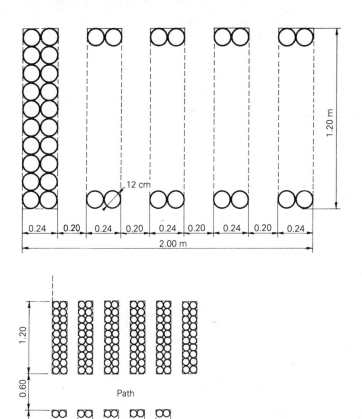

Fig 12 *Arrangement of the bags of soil for seed sown in the nursery*

The water used may contain a pre-emergence herbicide, based on Diuron, for example (15 g of commercial product in a concentration of 80 per cent in 100 litres of water would be adequate for 2500 bags).

Raising young seedlings

The seedlings emerge six to eight days after sowing. Regular watering has to be carried out at least once a day at the end of the day, preferably using a watering can (approximately 15 litres of water per 3 m² of bed). Watering can be reduced to every two days when the seedling has fully emerged, i.e. approximately 15 days after sowing, or when it rains. The soil in the bags should always be moist.

Apart from regular weeding, spraying is also necessary to combat predator insects and diseases, the most damaging of which are:
- defoliating caterpillars (*Earias* or *Anomis*). Treated by spraying the foliage thoroughly with a solution of acephate (Orthene 50 WP: 50 g of commercial product per 10 litres of water).
- the biting-sucking insects such as the psyllids (*Tyora tessmanni*). Plants should be sprayed every two months with a solution of acephate (Orthene 50 WP: 50 g of commercial product per 10 litres of water).
- damping off due to *Phytophthora* (the symptoms of which are a necrosis from the top downwards, then a general appearance of having been burnt), treated by spraying with metalaxyl (Ridomil '25' at 200 g/hl, i.e. grams per hundred litres, for example), or that due to *Rhizoctonia* (necrosis developing at the base of the plant at the collar), treated with a benomyl or oxyquinoline-based soil fungicide.

Snails also have to be eliminated by spreading molluscicides around the beds. Similarly, rodents and other animals have to be repelled by an enclosure made of wood or a fine-meshed metallic trellis part of which is sunk into the ground.

Finally, at least one month before planting out in the field, the plants must become accustomed to more intense light by progressively removing the shading from the nursery.

5 Establishing a cocoa plantation

Choosing the site

First it is important to have information on total rainfall and distribution per annum and an accurate assessment of the physical characteristics of the soil in which the cocoa trees are to be established. If rainfall is too low (1300–1500 mm per annum), excessively sandy or permeable soils should be avoided. If the rainfall is too high (2000 mm), a sandy soil can be used if there is no risk of a long dry-season. A sandy-clay loam soil is, however, always preferable.

There must be easy access to the site selected and it must be as close as possible to where the plantation manager lives. Its slope must not be greater than 10 per cent so as to enable cultural operations and harvesting to be carried out easily.

Denuded hilltops should usually be avoided, as should valleys which may have soils containing too much clay or are too sandy, also soils which are liable to flooding or which have poor water retention properties.

Forest land with a deep, sandy, clay soil which is as rich as possible in minerals is the most suitable. The minimum acceptable depth is 1.5 m, with a humus-bearing layer of at least 15 centimetres. The temporary or permanent water table must be more than 1.5 m below the surface.

The depth and texture of the soil must be checked visually. To achieve this, it is recommended that at least four 1 m³ trenches be dug for each hectare of land to be used.

Finally, a careful examination of the site will enable any land to be discounted which could favour the development of root rot. It is advisable to find out which crops have grown there before and to examine any dead or dying trees. An incision with a machete to the base of the tree will reveal whether there is any root rot in the form of white strips, smelling strongly of fungus, under the bark and inside the wood.

Land preparation

The planter must be aware of the fact that preparing the land for planting cocoa trees takes a great deal of time. Well-organised preparation will be beneficial to the cocoa trees' development and the time spent on this preparation will be turned to advantage all the more rapidly. Hasty preparation sooner or later leads to many problems which are often difficult and always costly to solve.

The land must be prepared at least one year before the cocoa seedlings are planted out. The preparation should involve the land being cleared, with shading being arranged so as to be ready to shelter the young plants when they leave the nursery.

The way to produce conditions most favourable to the subsequent development of the plantation is to replace the shade by establishing shade trees after felling the forest completely, even if this method is more costly and takes longer than simply clearing some of the undergrowth. The following steps should then be taken.

Complete clearance of the undergrowth

This should be carried out in the dry season and will allow free access to all the land. It will then be possible, depending on the topography of the site, to mark out the inspection and maintenance paths and, if the climate makes it necessary to do so, to mark out the network of drainage or irrigation channels. An inspection path suitable for motorised vehicles should be provided every fifty metres.

Felling forest trees

Large forest trees should also be felled in the dry season. The felled trees are sawn into lengths which are easy to handle. The pieces of trunk and branches are then piled up into wind-rows, together with all the vegetation generated by the land clearance. Gathering this vegetation into wind-rows must be carried out by hand or by using a winch. Mechanical vehicles (such as tractors and bulldozers) for felling and placing in wind-rows can be used on deep soils (alluvial soils, for example), but not on soils which have only a thin layer of humus.

Extraction of the stumps

It is necessary to remove as many of the tree stumps and roots as possible, otherwise they will be potential sites for root rot and could shelter rodents and other predators. Furthermore, the stumps will always be in the way of cultural operations. Stumps of large trees should be removed by chain-sawing and by progressively loosening the roots. Excessive disturbance of the soil should be avoided. This is a lengthy operation which can only be carried out by mechanical means if the soil is deep enough. In this case, the small trees and bushes should be pulled up, preferably by lifting rather than by pushing.

Staking out the land

The land should be roughly staked out, taking into account the direction of and distance between the future planting rows, both for the creation of the wind-rows after the forest has been felled and for the planting of temporary or permanent shading.

Burning the wind-rows

Burning is most effective in the dry season. This practice of burning must be limited solely to the wind-rows, the main aim being to clear as much of the land as possible without waiting for the cut wood to decompose naturally, which sometimes takes a very long time (more than twenty years for some hardwoods).

Establishing essential temporary shading

This must follow closely on the burning of the wind-rows so that the exposed soil does not become degraded by direct exposure to the sun or rain or infested with weeds.

Temporary shading is usually provided by food crops which give the planter an initial return on the land prepared for the plantation. Bananas are the most frequently-used crop, particularly the plaintain banana. These are planted at the same spacing as the cocoa trees and provide satisfactory shading six to nine months after planting.

Other food crops are used such as taro (*Xanthosoma* spp. and *Colocasia* spp.), particularly in West Africa, pigeon pea (*Cajanus cajan*), papaw (*Carica papaya*), cassava (*Manihot esculenta*) or the castor-oil plant (*Ricinus communis*). Fast-growing trees such as the *Gliricidia* or some erythrinae (*Erythrina lithosperma* in particular) are used in several countries as temporary shading.

> *Temporary shading, apart from its fundamental role of protecting the young cocoa trees from direct exposure to the light, also acts as a windbreak. The trees used as shading can therefore sometimes be planted in windbreak hedges.*

Protecting the ground with cover plants or food crops

Ground-cover plants, sown immediately after the burning, can be either erect, such as *Flemingia macrophylla*, which makes fine inter-row hedges, the foliage of which, after pruning, provides an ideal mulch for the soil, or more spreading species such as *Crotalaria* spp. or *Centrosema pubescens*. The latter has to be kept under control by regular cutting back. In general, it is relatively difficult and expensive to establish and keep vigorously growing plants adequately under control.

However, in the early stages of the preparation of the plantation, it is possible to intercrop with a variety of food crops, rice, peanuts, niébé, gumbo or maize.

Establishing permanent shading

(a) Permanent shading by retaining forest plants
This is shading intended to form the canopy over the adult plantation. It is recommended if a complete replanting of shade trees cannot be undertaken by the planter. In Africa, permanent shading generally consists of species being retained as the forest is being felled. In this case it is necessary to select suitable trees because many of them are known to have an unfavourable effect on cocoa tree development. This applies, for example, to *Piptadeniastrum africanum*, *Pentaclethra macrophylla* and all types of cola tree (*Cola* spp.).

Fig 13   *Cocoa trees under coconut trees* (Photo: G. Mossu)

In the same way, all trees which are known to be hosts for insects or diseases likely to attack the cocoa tree must be eliminated. Examples of these are all the Sterculiaceae and all the Bombacaceae (the silk-cotton tree, for example).

(b) Specially-planted permanent shading

In most of the producing countries of America, Asia or Oceania, permanent shading is provided by specially-planted trees, most commonly *Leucaena leucocephala*, *Gliricidia sepium*, various Erythrinae or even Albizia species. Finally, other industrial crops are often used as permanent shading for the cocoa tree, the coconut palm in particular.

The coconut palm

Intercropping cocoa trees with coconut palms is a common practice in South East Asia. It is most commonly adopted where the soil in the coconut groves is suitable for growing cocoa trees, such as the volcanic soils of Vanuatu, the clay soils of Malaysia or the alluvial soils of the island of Mindanao in the Philippines. Several different planting patterns are used in different countries for planting cocoa trees under coconut palms. If the coconut palms are spaced 9 m apart in a triangle, as is generally the case, one of the simplest patterns consists of planting the cocoa trees 3 m apart in a triangle. In order to allow maintenance work and harvesting of the two crops to take place, it is useful to leave one coconut tree inter-row out of three unplanted.

A well-maintained coconut plantation provides an excellent environment for intercropping with cocoa.

Other industrial crops

There have been several attempts to intercrop cocoa trees with Hevea oil palms, areca palm (*Areca catechu*) or even with nutmeg (*Myristica fragrans*). When planted at their normal density, all these plants produce too much shade for the cocoa trees, which develop as far as vegetation is concerned, but produce very few pods.

Regeneration of old cocoa plantations

Replanting young cocoa trees in an adult cocoa plantation has been tried many times more or less successfully. Before starting, it is essential to look into the causes for the decline of the plantation which is to be replanted. In particular, the planter should make sure that simple agricultural techniques alone will not suffice to restore the plantation. If replanting is to go ahead, the nature of the canopy should make it possible to select the method to be adopted.

If the permanent shading of the plantation is satisfactory and can be retained, the young cocoa trees can be planted in the inter-rows of the old plantation, taking care, however, to cut the lateral roots of the old trees away completely all around the young plants. Where there are gaps in the canopy, banana trees could complete the shading. When the trees of the young plantation are established, the old cocoa trees, which have continued to provide an income, are gradually removed.

If the permanent shading of the old plantation is non-existent, or is as poor as the plantation itself, it will often be better to fell all the shade trees, pull up the stumps and replant, after planting appropriate temporary shade trees using plantain bananas, for example.

The main ways of regenerating cocoa plantations are by:

- total replanting: after clear-felling, establish a new plantation under temporary shading. This is a costly process, but is generally profitable for small areas.
- partial replanting: part of the cocoa plantation is replaced each year in blocks or strips. This means that income can continue to be generated during the regeneration process.
- selective replanting: replacement tree by tree. This is not very effective. The young plants spread around the plantation are exposed to the existing diseases and insects infecting the old trees.
- replanting between old cocoa trees: temporary shading is provided by the old cocoa trees and they continue to provide income during the regeneration process, but plant health problems must be taken into consideration which may rule out this option.
- regeneration from stumps: this method consists of selecting a vigorous shoot from among all the shoots which appear on the stump of the old cocoa tree. This can be cut back completely from the beginning of the operation; this encourages shoots to develop, or, at the latest after the jorquette of the shoot selected has formed. In this case, temporary shading is guaranteed by the mother plant.
- regeneration by grafting: after cutting back the old cocoa trees, a bud taken from selected material is grafted on to one of the shoots which have appeared. This is a high-risk method which can only be carried out in countries where the labour is skilled. Grafting is also possible directly on to the trunk of the cocoa tree which is to be replaced. In this case, the tree is not cut back until the graft has taken and starts to develop.

Planting

Planting cocoa is a simple operation. Nevertheless, it should be borne in mind that a cocoa plantation should be productive for a minimum of 25 years and that it is also often very difficult, both technically and economically, to rehabilitate a poorly-established plantation after a number of years. The long-term investment which a plantation represents therefore means that great care must be taken at the different stages in this operation.

Pattern and density of planting

Several planting patterns can be adopted: in squares, in a staggered arrangement, in an equilateral triangle, in an isoceles triangle, etc., but the simplest and also the most commonly adopted pattern is to arrange them in equidistant rows, which greatly facilitates maintenance, supervisory and plant health work.

Density of planting depends on various factors and, in particular, on how vigorous the trees are, the nature of the planting material used, the level of shading, the soil and the climate. The density should be modified if the environmental conditions depart from the optimum conditions described below (cf. Chapter 7). For example, if the light is intense, it is an advantage to plant the cocoa trees more closely together.

The genetic origin of the cocoa trees must be taken into consideration. Cocoa trees belonging to the Lower Amazonian Forastero group (Amelonado, for example) grow rather more slowly and the development of the jorquette is characteristically delayed. These types can therefore be planted closer together than the Upper Amazonian Forasteros or hybrid cocoa trees which are more precocious.

In general, the following principles can be adhered to:
- a spacing of 3 m × 2 m (1666 trees/hectare) is most suitable in areas where the rainfall does not exceed 1500 mm per annum,
- a spacing of 3 m × 3.5 m (952 trees/hectare) is recommended for slightly wetter areas, on good soils or on well-drained soils regularly receiving applications of fertiliser,
- a high density of planting (1600 to 2000 trees/hectare) is preferable on very well-drained soils, if no mineral fertiliser is applied,
- for particular environmental, maintenance and tending conditions, in certain countries (Malaysia, for example), there are arrangements for very high-density planting, of 5000 trees/hectare or more, usually planted in double rows at a spacing of 1 m × 1 m with 3 m between the double rows.

Fig 14 *Original planting technique: very high-density planting in double rows*
(Photo: G. Mossu)

Marking-out – preparing the planting holes

When the land and the shading have been correctly prepared, the final
marking-out can be carried out based on the planting pattern and density
which have been selected. Stakes are used to mark each planting hole.

Marking-out is generally followed immediately by preparing the planting holes. This consists of digging a 40 cm cube-shaped hole at each
stake. It is a process which is very often neglected or even completely
left out because of claims that it is a waste of time and labour. Unless
there are exceptionally good soil conditions, where a simple hole with
the dimensions of the seedling bag can be made at the time of planting
out, hole-making is, in fact, a very important operation which helps the
plants to become rapidly established. It loosens the soil, improves moisture levels and enables any root debris, a possible source of root rot, to
be removed along with concretions and pebbles which are likely to be
damaging to the development of the root system.

The holes, the walls of which must not dry out through prolonged
contact with the air, are immediately filled in again with topsoil. This
will provide an excellent substrate for the first stages of root growth.
Apart from the addition of calcium or dolomitic compounds, mixing
fertilisers in with the hole-filling soil is not recommended. On the other
hand, the addition of manure or well-decomposed compost may be beneficial in eroded soils, or soils which are deficient in organic matter.

On land which has been completely cleared of stumps, hole-making
can be replaced by the use of a subsoiling plough which passes along the

planting row; this should be followed by a disc-harrow to level the ridges. The practice of subsoiling is not recommended for heavy and clay soils.

Hole-making is followed by the application of a complete herbicide, which is sprayed along the planting rows. The best formulae are based on glyphosate (Roundup at 360 g/l of glyphosate: 4 litres of commercial product per hectare), or paraquat (Gramoxone at 200 g/l of paraquat: 1.5 litres of commercial product per hectare). These herbicides are diluted in 300 litres of water per hectare, which is equivalent to using a 10 litre spray for a 100 metre row.

Planting the seedlings

Planting must take place as soon as the rainy season is well established and as long as possible before the following dry season. The day before planting out, the most vigorous seedlings, all the leaves of which should be dark green, are selected in the nursery. Small, twisted or straggly plants, as well as seedlings in the middle of flushing, are kept in the nursery. The plants which are selected for planting are watered copiously, transported to the field and placed in a cool, shaded place. Planting must take place in the early hours of the morning and, if necessary, at the end of the afternoon, but never during the hottest part of the day. Each plant is subjected to the following planting procedure:

- open the planting hole up to the dimensions of the pot and where the stake is;
- remove the bottom portion of the bag (plastic and soil) approximately 2 cm from the base. The taproot, which is often coiled round at this point, is therefore cut. It will grow again vertically into the earth by growing a new orthotropic apex;
- cut the plastic of the bag, from top to bottom;
- place the seedling in the hole, remove the plastic and take care that the collar of the young plant remains at ground level;
- gradually refill the hole around the root ball with soil until the hole is full, making sure it is firm by treading. Loose planting always results later in a high percentage of dead plants;
- place a protective mulch of leaves and twigs all around the cocoa plant, but prevent it from coming into contact with the collar;
- make a shelter, of palm fronds, for example, to give additional shade and a little protection (from winds and insects) to the cocoa plants until they are adequately established; this will take approximately two months.

Under normal rainfall conditions the overall percentage of losses in terms of failure to take is 3 per cent to 5 per cent in a well-run plantation. The plants which fail will be replaced in the following rainy season with surplus cocoa plants held in the nursery.

It is, however, possible to keep the seedlings in the nursery for longer, cutting back the main stem to 30 cm from soil level when they are five to six months old, and, at the same time, pruning away the roots which have grown outside the bag. This method allows the plants to be kept for a further year and poses no problems when it comes to the plants 'taking' in the field.

Plantation maintenance

Maintaining the plantation does not present any particular technical problems. It is, however, recommended that the planter should always be vigilant, particularly during the early years of the young plants' development. Maintenance also implies a rational management of time and means. The planter's calendar must take account of the ongoing operations which can be carried out in rotation, such as adjusting the shading, weeding and sucker removal, but also one-off and unavoidable tasks for which all the available labour can be used. This applies to pest and disease control and harvesting, for example.

The main maintenance operations are as described in the rest of this chapter.

Adjusting the shading

Adjusting the shading is, without a doubt, one of the most important elements in maintaining a growing plantation.

During the first year dense shade, allowing through 50 per cent of the total light, is necessary. As we have seen, this can be achieved by planted or adjusted overhead shading, often supplemented by interplanted hedges from a natural regrowth or from sown erect cover plants. These interplanted hedges will be trimmed regularly to make them less dense and will gradually be removed altogether.

The temporary shading must then be progressively reduced to allow through 50 per cent to 75 per cent of the total light (or even 100 per cent if it is considered that the plantation can develop satisfactorily without shading). This progressive reduction (or elimination) of the shading should only take place when the cocoa trees are fully developed and when their branches overlap to form an unbroken canopy.

Soil maintenance

Soil maintenance consists mainly of removing casual weeds manually or by using herbicides. The latter are very effective in young plantations. The various products used should, however, be alternated according to their mechanism of action (contact or translocation herbicides). The most commonly-used herbicides are paraquat (Gramoxone) at a rate of 1.5 litre of commercial product containing 200 g/l of paraquat in 300 litres of water per hectare, and glyphosate (Roundup) at 3 litres of commercial product to which 10 kg of ammonium sulphate has been added in 300 litres of water per hectare.

Some resistant dicotyledons have to be specially treated with a selective, hormone-based weedkiller such as 2, 4-D at a rate of 750 g of acid equivalent per hectolitre.

However, at the field's edge, on inspection paths and, if there are any unshaded patches, paraquat, glyphosate or ametryne and simazine-based compounds are very effective.

> *The colonisation of the upper layers of the soil by lateral roots prevents the soil being ploughed up or simply being weeded, whatever the age of the plantation. This means that any interplanted food crops can only be retained for a very short time, two years at the most, after the young cocoa trees have been planted out.*
>
> *In a well-established adult plantation, the carpet of dead leaves and the shading of the ground by the cocoa trees normally eliminates any weed growth.*

Pruning cocoa trees

The cocoa tree requires very little maintenance. This mainly consists of removing the suckers as well as any dead or diseased wood and, if necessary, opening up the canopy by removing some of the ends of the drooping secondary branches when these are too dense.

During the young plant's development it frequently happens, particularly on plantations which enjoy good lighting conditions, that the jorquette of the tree forms at too low a level. This is a serious obstacle to maintenance or harvesting operations and it is usually necessary to allow a vigorous sucker to develop, from just below the jorquette, to form a second trunk and a second jorquette.

> *The cocoa tree should not be subjected to the type of pruning normally carried out on fruit trees.*

Mineral fertilisers

Fertilisers are rarely used in cocoa-growing. The effect of fertilisers on the productivity of these trees depends largely on the cultural and light conditions. Under permanent shading, the use of fertilisers is rarely profitable. On the other hand, for cocoa trees grown in full sunshine, fertilisers are effective if, and only if, all the cultural conditions are correctly carried out and, in particular, if the rainfall is well-distributed and exceeds 1300 mm/year. The water supply may depend on rain or sprinkler irrigation may be used.

The determination of the optimum composition and rates of application of fertilisers for use on cocoa is carried out by soil analysis, a method based on an understanding of the anionic and cationic balances which promote the growth and productivity of the cocoa plants (Chapter 7).

Simple calculations based on an analysis of composite soil samples, representative of the plantation itself or of a homogeneous geographical area, enable the quantities of fertiliser to be determined which are needed to attain these balances in the topsoil while compensating for the elements exported at each harvest.

The harvesting of one tonne of commercial cocoa (without returning the pod residues) represents an average export of 45 kg nitrogen (N), 65 kg potassium (K_2O), 13 kg magnesium (MgO), 13 kg phosphorus (P_2O_5) and 10 kg calcium (CaO).

Fertilisers are spread over a distance between 0 and 30 cm around the base of the cocoa tree upon planting out, at a distance of between 30 and 60 cm two or three years after planting and between 60 and 100 cm from the fourth year. The period over which a specific fertiliser formula can be used is limited. To maintain the required balances, it is vital to regularly check the soil reserves by carrying out new analyses every three or four years.

> *Fertiliser application must be carried out from the year of planting if the soil is low in minerals and one year after planting on richer soils.*

Symptoms of some mineral deficiencies in cocoa

The symptoms which are characteristic of the different deficiencies, which are evident mainly from the appearance of the leaves, are well known and can help to diagnose qualitatively any mineral imbalances in the soil. Of these, the most frequently encountered are:

- nitrogen deficiency: the leaves are smaller than usual, with a discoloration and yellowing of the lamina and of the veins;
- potassium deficiency: olive-green discoloration of the leaf margins followed by necrosis;
- magnesium deficiency: yellow discoloration here and there from the central vein of the leaf;
- zinc deficiency: typical malformation of the very narrow leaves, which

are elongated and often furled in a sickle shape with very marked aberrant veining;

- boron deficiency: the young leaves are abnormally elongated, often shrivelled, are furled at the end and have yellow patches at the point of curvature. At the more advanced stage, the tips of the leaves are necrosed from the curvature downwards. Moreover, in the pods the seeds clump together without pulp.

Plant protection

In very general terms, in addition to the usual cultural procedures, plantation maintenance often involves costly plant protection treatments to combat damage caused by pests or diseases which prevail with greater or lesser intensity depending on the country and the growing conditions.

More than five hundred different species of insect pest have been recorded on the cocoa tree. Fortunately, in most cases the effect of these pests on productivity is negligible and does not justify the application of costly agro-chemicals. On the other hand, there are some major pests, both in the nursery and in the plantation, which sometimes cause considerable damage which could be limited substantially, or even totally avoided, by timely treatments. The effectiveness of the treatment clearly depends on the severity and rate of development of the infestation and on whether the biological cycle of the pest is known or not, but it also depends, above all, on the choice of an appropriate insecticide. An error of choice may work against the desired effect.

As for the diseases, they can directly affect the harvest by attacking the pods or indirectly by weakening and killing the trees. In certain sites several diseases can be controlled economically by simple cultural methods, in particular by adjusting the level of shading or by improving soil drainage. However, it is often necessary to resort to chemical treatments.

Overall it is estimated that losses of commercial cocoa amount to 21 per cent of potential world production because of diseases and to 25 per cent of this same potential because of insect pests. Together, the combined activities of diseases and pests mean that only 54 per cent of the potential total production reaches the world market.

Fig 15 *Left*   Distantiella theobromae (After E. M. Lavabre)
Fig 16 *Right*   *Mirids:* Sahlbergella singularis (Photo: M. Boulard)

Fig 17   *Mirids: damage caused by mirids on young cocoa trees* (Photo: G. Mossu)

Key to pests and diseases

By the use of Tables 3 and 4 the planter can identify most of the diseases and pests from the symptoms observed on the pods and on the rest of the plant. The planter will also find here the main control methods.

Fig 18   *Mirids:* Helopeltis (Photo: M. Boulard)

Fig 19   *Mirids: damage to pods*
(Photo: M. Boulard)

In these tables the name of the disease or pest is followed, in brackets, by the symbols:

FU,   for a fungus
IN,   for insects
VIR,   for a virus
MAM, for mammals
?   cause not identified

51

Table 3 *Damage to the pods*
A – Fruit not deformed

Symptoms	Location information	Name and causal agent	Treatment to be applied
A1 The pods have one or more brown spots which are hard to the touch and are covered with a creamy white spore-carrying coating. These spots ultimately cover the entire pod which, when it ages, dries and turns grey. The brown spot has a characteristic fishy odour. Sometimes considerable losses at harvesting (up to 95 per cent). This disease can also attack the leaves, the branches or the roots where it produces canker. In the nursery, young plants may turn brown from the terminal bud downwards and at the leaf margins. Very high risk of spreading.	Universal. Dependent on the rainy season. Favours plantations with limited air movement, particularly if the climate is very humid and the temperature is relatively low (15–18°C).	PHYTOPHTHORA POD ROT (FU) OR BLACK POD • *Phytophthora* sp.	*Preventative:* • Harvest the affected pods and cherelles. • Open the pods outside the plantation. • Reduce the ambient humidity by improved aeration and adjusting the shading. *Chemical:* • Spray the pods with a water-soluble fungicide at 21-day intervals using a metalaxyl-copper mixture (Ridomil 'plus', at 33 g commercial product to 10 litres of water).

	Symptoms	Conditions	Name	Control measures
A2	The pods initially develop brown marks which are soft to the touch. They then become completely black as if they were covered with a sooty powder. Relatively few losses in general, but may in some cases be quite high (more than 50 per cent).	Universal. Signs of weak growth following wounds, insect bites or primary attack by *Phytophthora* spp. or other pathogens.	CHARCOAL POD ROT (FU) • *Botryodiplodia theobromae*	• Sanitary harvesting • In severe cases, a fungicide such as iprodione (Rovral) at 2000 ppm should be sprayed every month for six months.
A3	The pods are speckled with small, individual, hardened blackish marks. The fruit itself is not badly affected but these symptoms are associated with other far more serious symptoms for the growing tree (see D.1). Very serious losses due to entire plots dying.	Universal. Insects which favour gaps in the canopy of foliage or in the shade above the tree, a large number of pods, moderate but uniformly distributed rainfall and a forest environment rich in plant hosts for the insects (*Sterculiaceae* and *Bombacaceae*)	MIRIDS (IN) • *Helopeltis* (Universal) • *Sahlbergella* sp. • *Distantiella* sp. • *Odoniella* • *Bryocoropsis* (Africa) • *Monalonion* (America) • *Pseudodoniella* (Pacific) • *Platyngomiriodes* (South East Asia) • *Boxiopsis* (Madagascar)	• Remove all the suckers by pruning • Remove the host trees (*Sterculiaceae, Bombacaceae* . . .) • Retain the overhead shading and foliage of the cocoa • Two applications, with 28 days between them, when the populations start to increase, with alternate forms of insecticides (lindane at 300 g active ingredient/ha in approximately 40 litres of water in a spray or in 2 litres of diesel oil by thermonebulisation. Alternate the lindane with: Thiodan (250 g a.i./ha – Unden (250 g a.i./ha) – Basudin (410 g a.i./ha).

Symptoms	Location information	Name and causal agent	Treatment to be applied
A4 As with *Phytophthora*, the pods have one or more brown marks. However, unlike *Phytophthora*, the marks have a circular depression in the middle in which a crust of spores quickly grows which rapidly turns pink.	Central and West Africa. It is mainly a wound parasite.	MEALY POD ROT (FU) • *Trachysphaera fructigena*	• The same precautions as for *Phytophthora*. • Frequent and regular harvesting together with rodent control.
A5 The pods appear to be normal, but, where infestation is heavy, they appear to have ripened prematurely. Careful observation reveals tiny holes. The seeds do not develop, the pulp is compact and hard. The pods on the branches are more seriously affected than those on the trunk. Serious losses at harvesting.	South East Asia – Pacific Insects favoured by the proximity of host plants, in particular rambutan (*Nephelium*) and cola (*Cola nitida*).	POD BORER (IN) • *Conopomorpha cramerella*	Growing method: • Remove cherelles and pods which are longer than 5 cm in periods when harvesting is light (interrupting the insect's life cycle). All these pods and cherelles should be burnt. • Burn the pod husks after harvesting or isolate them in plastic bags. • Covering the developing pods with plastic is the best remedy, but is difficult in the case of pods formed on the upper branches.

A6 Pods develop more or less woody brownish areas from which branched brown galleries form on the surface. The parenchyma tends to split. The symptoms often look serious, but no severe damage is done to the fruit.	Central and South America Africa	POD BORING CATERPILLAR (IN) • *Marmara* sp.	• No special treatment.
A7 Pods develop hollows approx. 0.5 cm in diameter, then are covered with a pink powder. Generally little damage to the fruit, but accompanying symptoms are far more serious for the young leaves, including repeated defoliations of the young shoots leading to the opening of the axillary buds. Disordered growth.	Universal. Favoured by high humidity.	ANTHRACNOSE (FU) • *Colletotrichum* sp.	• Sanitary pruning and harvesting.

A8 Pile of wood dust and debris caught by thread of silk at the base of the pod in contact with the trunk or when two pods touch. This debris comes from openings and galleries excavated in the pericarp of the fruit, without affecting the seeds, but the damage encourages rot and other fungal diseases to develop. Many pods lost.	Africa. When the pods ripen.	POD-FEEDING MOTH (IN) • *Characoma stictigrapta*	• In the event of serious damage, use an insecticide based on deltamethrin (Decis).
A9 The stalk or base of the pod has dark brown, reddish-brown or completely white, floury and oval warty clusters. These clusters may even form sleeves around the stalk of the pod. Pods sometimes damaged but symptoms are more serious in other parts of the plant (young shoots in particular), where the leaves may be curled in the plane of the lamina or blistered and shrivelled and there is the risk of the transmission of a serious disease (virus diseases in particular).	Universal. These symptoms are very generally associated with the simultaneous presence of large numbers of ants, which build their nests of earth and vegetable debris between the stalk of the pod and the branch of the trunk.	MEALYBUG (IN) • *Planococcus* sp. • *Ferrisia* sp. (Universal) • *Stictococcus* sp. (Africa)	• Treatment with a systemic insecticide such as monocrotophos (1 volume/1 volume of water). Paint it on to the trunk or inject it into the base of the trunk. • Treat by spraying acephate-based foliage insecticides in doses of 500 g active ingredient/hectare. • Eliminate the host trees (*Sterculiaceae, Bombacaceae, Tiliaceae* and *Malvaceae*, for example).

Fig 20 *Mealybugs:*
Ferrisia on a young shoot
(Photo: M. Boulard)

Fig 21 *Bug: Bathycoelia on a pod*
(Photo: M. Boulard)

Fig 22 *Psyllid:* Tyora *larvae on a*
branch (Photo: M. Boulard)

Fig 23 *Stem borer:*
Tragocephala *sp.*
(Photo: M. Boulard)

Fig 24 *Defoliating lepidoptera:* Earias
sp. damage (Photo: Nguyen-Ban)

Fig 25 *Thrips: defoliation by* Solenothrips
(Photo: N. Coulibaly)

Fig 26 *Rodents: damaged pods* (Photo:
E. M. Lavabre)

B – Deformed pods

Symptoms	Location information	Name and causal agent	Treatment to be applied
B1 The adult pods develop brown marks with sinuous edges covered with a mother-of-pearl type of spore-carrying coating, usually before ripening. This same type of mark also appears on young cherelles on which the first visible symptom of the disease is a swelling of variable shape. Inside the fruit, the seeds are reduced to a more or less liquefied brown mass, even before the symptoms have appeared externally. The pods dry on the tree and, at the slightest contact or gust of wind, give off a cloud of spores. Serious losses of up to more than 70 per cent at harvesting. Very high risk of infection.	South and Central America. High humidity favours spore germination. The spores are dispersed very widely by the wind.	MONILOPHTHORA POD ROT (FU) or wet rot. • *Monilophthora roreri*	*Prevention:* • Remove the mummified pods before new fruits develop. • Weekly harvesting of the affected cherelles and pods. • Reduce the ambient humidity by aeration and adjust the level of shading. • Reduce self-shading by pruning drooping branches twice a year. *Chemical:* • Apply copper oxide every week in the flowering-fruit formation period, then every 10 to 12 days.

Fig 28 *Cocoa fruiting* (Photo: G. Mossu)

Fig 29 *A fine adult cocoa tree in production* (Photo: G. Mossu)

Fig 30   *Leaf symptoms of a number of mineral deficiencies* (Photo: A. Loué)
1. Normal leaf, 2. Nitrogen deficiency, 3. Potassium deficiency, 4. Magnesium deficiency,
5. Zinc deficiency, 6. Boron deficiency

Fig 31   *Mirids: insecticidal treatment by thermonebulisation* (Photo: G. Mossu)

Fig 32   *Black pod: damage due to*
Phytophthora *sp. on pods*
(Photo: G. Blaha)

Fig 34   *Moniliophthora: deformed
young cherelle* (Photo: O.
Trocmé)

Fig 33   *Witches' broom disease: affected
branch* (Photo: R. A. Muller)

Fig 35 *Above   Swollen shoot:
mosaic type of symptom on a leaf*
(Photo: M. Partiot

Fig 36 *Right   Swollen shoot:
symptom on a branch*
(Photo: M. Partiot)

Fig 37   *Root rot: affected tree* (Photo: G. Mossu)

Fig 38   *Preparing commercial cocoa: basket fermentation* (Photo: G. Mossu)

Fig 39   *Preparing commercial cocoa: drying 'autobus' in an industrial plantation* (Photo: G. Mossu)

Fig 40   *Preparing commercial cocoa: movable roof-dryers in an industrial plantation* (Photo: G. Mossu)

Fig 41   *Preparing commercial cocoa: wood oven of the 'Samoa' type* (Photo: G. Mossu)

Fig 42 *Preparing commercial cocoa: hot-air oven equipped with a movable rake for stirring the beans* (Photo: G. Mossu)

Fig 43 *Storing commercial cocoa* (Photo: G. Mossu)

B2 The pods remain small, assuming an atrophied shape. They then dry and rot on the tree without ever reaching maturity. The symptoms on the pods are accompanied by far more noticeable symptoms on the branches, young shoots and flower cushions, where the disease produces hypertropic growth of the buds. These then develop into a 'witches' broom' with a great increase in the proliferation of branches. These brooms die after 5 to 6 weeks and carry small pink fungi on the dead wood. Considerable loss of pods (up to 90 per cent) and a very high risk of contamination and spread of the infection.	South America Central America The West Indies	WITCHES BROOM • *Crinipellis perniciosa* • Keep the plantation shaded (this limits fluctuations in temperature, humidity and air movements which may favour it). • Routine pruning: cutting off all diseased branches and cutting out diseased flower cushions. All the diseased tissues must be burnt. • No conclusive results with fungicides as yet.

Description	Distribution	Pest / Pathogen	Control
B3 Pods clearly divided into two parts. One part, which is more yellowish (apical or basal), appears to be atrophied and sometimes necrosed. The seeds are empty and brown. Fairly serious damage (more than 20 per cent losses in Ghana, for example).	West Africa South East Asia Favoured by an abundance of pods throughout the year (e.g. High Amazonian hybrids) and by openings in the canopy of the trees. Also favoured by the dry season.	SHIELD BUGS AND COREID BUGS • *Bathycoelia* sp. • *Pseudotheraptus* sp. • *Amblypelta theobromae*	• The same chemical treatment as for mirids. • Interrupt the life cycle of the insect by totally removing all the fruits once a year when harvesting is not being carried out.
B4 Pods have numerous warty protuberances which become black and soft when the fruit ripens. Losses upon harvesting only affect overripe pods. Currently relatively few losses, but the disease is rapidly spreading to different areas.	West Africa	WARTY POD DISEASE • Primary pathogen not identified (probably an insect) • Systematic secondary infection by pod rot (*Botryodiplodia theobromae*)	• Harvest the pods when ripe. • Chemical treatment is under consideration.
B5 Pods gnawed in such a way as to make a more or less circular opening and completely emptied of their contents. The pods dry and become mummified on the tree. Sometimes considerable damage, affecting up to 90 per cent of the pods.	Universal Ripe pods particularly affected but may happen to fruit at any age, particularly in isolated, rarely inspected and poorly maintained plots within dense shading and dense weed growth.	RODENTS (MAM) • Many species of rodent (rats, squirrels, etc).	• Clean the plantations, separate the plantation from the forest by roads, frequent inspections. • Organised hunts. • Traps. • Poisoning by using an anticoagulant (provided that there is no likelihood of the animal being subsequently eaten).

Table 4 *Damage to the vegetation*
C – Generalised damage leading to the death of the tree

Symptoms	Location information	Name and causal agent	Treatment to be applied
C1 The foliage withers suddenly and the leaves remain attached to the tree. No living shoots at the base of the tree which may have fallen. Whitish, yellowish or black, irregularly branching strands run along the roots. Vertical cracks and splits may appear on the bark of the trunk at the collar. In the first years after planting, the cocoa trees affected are always isolated. The disease then spreads in patches and can affect a great many trees in the plantation.	Universal. Occurs particularly in plantations where the stumps of large trees have been retained and have not been either extracted or poisoned.	ROOT ROT (FU) ● *Armillariella mellea* ● *Fomes* sp. ● *Rosellinia* sp.	● Extract or poison the stumps. ● Careful uprooting of the diseased trees with all their roots. ● Dig trenches around infected areas. ● Chemical treatment under consideration based on cyproconazole (Alto 100:10 ml of commercial product in 1 litre of water poured on to the base of each tree).

Symptoms	Location information	Name and causal agent	Treatment to be applied
C2 A considerable drop in production and abnormal morphogenesis apparent, in particular in the appearance of swellings in the wood of the branches, the roots and shoots. The very young leaves of the leaf shoots have a reticulated mosaic pattern and chlorosis along the veins. An affected adult cocoa tree can die in three years. Very serious losses and a very high risk of contagion.	West Africa Disease transmitted by mealybugs (see A9). Similar or less severe symptoms have also been recorded in many countries in other continents.	SWOLLEN SHOOT (VIR) Cocoa Swollen Shoot Virus (CSSV) • Various more or less aggressive forms of the virus have been identified in Ghana and in Togo.	• Control the mealybugs (see A9) by painting on or spraying insecticides. • Cut out the trees showing symptoms, as well as neighbouring trees which appear to be healthy. This must be done at an early stage.

Symptoms	Conditions	Cause	Treatment
C3 Poorly developed leaves, many ends of branches lose their leaves and often wither. Basal shoots develop. The ends of these shoots may become blackened and wither. Several branches may have the appearance of candelabra. Canker can be observed on the branches and in the forks of branches. Small white pustules can be seen in vertical lines on the dead wood. All the foliage withers from the top to the bottom, rapidly followed by the death of the tree.	Universal. Virtually always follows upon an attack by mirids and is generally aggravated by inadequate shading.	TRACHEOMYCOSIS due to CALONECTRIA (FU). • *Calonectria* sp. • *Colletotrichum* sp.	• Reinstatement of conditions favourable to the development of the trees. • Anti-mirid treatment (see A3). • Cut out the affected branches below the lesions.
C4 Some leaves wither but do not fall or curl. Withered shoots occur at the base of the tree and on the side where the foliage is diseased, with healthy shoots elsewhere. Sections of the tree die due to the disease progressing from the bottom to the top.	Universal. Favoured by intense evapotranspiration and also by inadequate shading.	TRACHEOMYCOSIS due to VERTICILLIUM (FU) • *Verticillium* sp.	• Reduce evapotranspiration by increasing the shading. • Remove the diseased branches. • Possibly rejuvenate the tree by developing a healthy shoot.

Symptoms	Location information	Name and causal agent	Treatment to be applied
C5 Yellowing, with characteristic green marks on some adult leaves in the centre of the branch, developing progressively towards the base and the ends of the branches. Beyond these leaves, towards the end of the branches, the axillary buds grow normally before dying. The disease slowly reaches all the branches and ultimately kills the tree, whatever its age. The wood has brown streaks running through it which become visible on tangential section. Sometimes very serious damage (up to 50 per cent of the trees). Very high risk of infection.	South East Asia. Favoured by the rainy seasons and by nocturnal winds. Young cocoa trees are very susceptible to this disease, which is often widespread in the nursery. Cocoa trees weakened by malnutrition or insect attack are also very susceptible.	VASCULAR STREAK DIEBACK (FU) (VSD) • *Oncobasidium theobromae*	• Plant the trees in such a way that they can grow vigorously and become well-developed. • Establish a wind-break. • If trees are severely affected, carry out general pruning of the branches of the affected tree. • Burn the pruned branches.

Pest / Disease	Distribution and notes	Symptoms	Control
CERATOCYSTIS WILT (FU) • *Ceratocystis fimbriata*	Central and South America South East Asia Has not been reported in Africa A disease associated with wounds made to the tree and particularly with attacks by insects of the Xyloborer type, which bore into the wood and play an important part in transmitting the disease.	C6 Sudden, rapid withering of part or all of the foliage, followed by necrosis and rapid death of the infected areas. The adult leaves drop and turn brown. Curled lengthwise, they remain attached to the tree. The damage may be very serious: up to 50 per cent of the trees may die.	• Prune and burn branches and dead trees which have been attacked. • Avoid damaging the tree during cultural operations. • Sterilise all pruning tools.
COCKCHAFER (IN) • *Camenta* sp. • *Phyllophaga* sp.	Africa. The West Indies.	C7 Death of the tree without any sign of external symptoms. The taproot of the tree or the large lateral roots are eaten away close to the collar by a large white grub.	• Luminous night traps. • Treatment of the soil with insecticide in granular form around the tree collar (Phorate, for example: 1 g active principle/tree).
DEFOLIATING BEETLES (IN) • *Adoretus versutus* • *Rutela lineola* • *Pseudotrochalus concolor* • *Apogonia derroni* • *Nodonota theobroma*, etc.	Universal.	C8 Severe defoliation of the tree. The young plants are particularly liable. The leaf shoots are systematically destroyed, as are the adult leaves. The tree eventually dies. Damage which may be severe.	• Protect the young plants individually by sheltering them under palm fronds, for example. • Treatment of the soil with insecticide in granule form directly below the foliage (Phorate, for example: 1 g active principle/tree).

D – Damage to the branches

Symptoms	Location information	Name and causal agent	Treatment to be applied
D1 Many withered branches still carry some dead leaves. No sign of holes, galleries or piles of sawdust. Green branches which remain alive have numerous elongated brown patches, which sometimes fuse lengthwise. Leaves turn brown from the veins outwards. The lignified wood has more or less furrowed depressions. Fruits, if there are any, are covered with scattered blackish brown spots.	Universal The withering and death of the branches (and sometimes of the entire tree) are generally due to secondary tracheomycosis (C3 and C4), introduced into the wood where mirids have sucked the sap.	MIRIDS (IN) See A3	See A3
D2 No dead leaves remain attached to the ends of the withered branches. Generally, there is a characteristic shortening of the internodes and development of the axillary buds, giving a characteristic fin-like appearance to the affected branches.	Africa Attacks leaf shoots in particular, both in the nursery and in the open field. Disrupts the growth of the young plants.	PSYLLIDS (IN) • *Tyora tessmanni*	• Insecticide belonging to the organophosphate family (parathion, monocrotophos or acephate).

Description	Name / Distribution	Control
Many thread-like flakes adhere to the ends of the young shoots or to the flowers and flower cushions, or even under the young leaves which become blistered or shrivelled. Sometimes very serious damage to young plants exposed to full light. Up to 70 per cent of the plants may be lost.		
D3 Withered branches, either isolated or in small groups, still bearing their dead leaves. Circular holes present from which sawdust falls, sometimes agglomerated in clusters. Galleries are excavated which may continue into larger branches. A circular annular incision may be visible towards the rib of the branch. The terminal bud withers.	STEM-BORERS and WOOD-BORING CATERPILLARS (IN) • *Tragocephala* sp. • *Glenea* sp. • *Steirastoma* sp. • *Xyleborus* sp. • *Erodiscus* sp. • *Pantorhytes* sp. • *Eulophonotus* sp. • *Zeuzera* sp. Universal Frequently seen on young plants which are exposed to full light and in neglected plantations.	• Cut the dead branches. • Use systemic insecticide if possible. • Maintain the plantation efficiently and leave a cleared area between the plantation and the neighbouring forest.
D4 On branches and trunk: hypertrophic development of the flower cushions on which either multiple small green 'buds' or a great many sterile flowers appear. A disease which has an impact on harvesting, but is rarely serious.	CUSHION GALL (FU) • *Calonectria rigidiuscula* (or *Fusarium decemcellulare*) America The West Indies Africa	• Paint the trunk with a systemic fungicide.

E – Damage to the leaves

Symptoms	Location information	Name and causal agent	Treatment to be applied
E1 Margins of the leaves have scorched areas, possibly blisters. Leaves are deformed: incised contours, the end aborted or necrosed. Sometimes very serious damage.	Universal Favoured by plants being exposed to full light or by the sudden removal of shading.	LEAF HOPPERRS (IN) • *Empoasca* sp. • *Affrocidens* sp. • *Chinaia* sp. • *Horiola* sp.	• Insecticide: Fenitrothion and formothion sprayed at a rate of 1000 g of active principle per hectare.
E2 The leaves are eaten and the terminal bud of the young shoots is aborted. It generally contains a grub. Growth is disrupted; excessive delays in the appearance of the jorquette, or, under unfavourable conditions, the plant may wither. Severe defoliation may sometimes occur in adult plantations.	Universal Favoured by plants being planted in full light or by the sudden removal of shading.	COCOA BOLLWORM • *Anomis* sp. • *Earias* sp. • *Tiracola* sp. • *Melanchroia* sp.	• Systemic insecticide to control the borers: monocrotophos. Contact insecticide for the defoliators: Endosupgan (250 g active principle/ hectare).
E3 The leaves become discoloured in patches, then turn yellow and rust coloured before they fall. Severe defoliation may sometimes occur in adult plantations.	Universal Frequent during periods of water shortage. Favoured by the sudden removal of shading.	THRIPS (IN) • *Selenothrips rubrocinctus*	• Insecticide: Fenithothion, diazinon, propoxur.
E4 Leaves eaten away and severely damaged, often only leaving the central vein intact. Sometimes very serious damage.	South America Central America West Indies	LEAF-CUTTING ANTS (IN) • *Atta* • *Acromyrmex*	• Destroy the nests by fumigation with methyl bromide or phosphoretted hydrogen (PH_3).

6 Harvesting and preparation of commercial cocoa

Commercial forms of cocoa are obtained after the harvested seeds have undergone fermentation and drying. The seeds are then referred to as cocoa beans and usually exported in this form to the user countries.

Harvesting the pods

From the pollination of the flowers up to the ripening of the fruits, the formation and development of the pods take, on average, five to six months. The pod then changes colour, with the green pods turning yellow and the red pods turning orange. The pods should be harvested when they are ripe. A ripe pod is particularly vulnerable to diseases (various types of rot in particular) and predators (rodents). Furthermore, if the pods are left on the tree for too long, the seeds will germinate, which makes them unsuitable for turning into commercial cocoa. However, it is even more serious to harvest the pods before they are ripe, because fermentation of the seeds in this case always produces a poor-quality cocoa which is low in aromatic compounds.

The pods are harvested by making a clean cut through the stalk with a well-sharpened blade. For pods high on the tree, a pruning-hook type of tool can be used with a handle at the end of a long pole. By pushing or pulling according to the position of the fruit, the upper and lower blades of the tool enable the stalk to be cut cleanly without damaging the branch which bears it.

> *Harvesting should be carried out at regular intervals of ten to fifteen days which, in any event, should never exceed three weeks.*
>
> *During harvesting, it is important not to damage the flower cushion which will produce the flowers and fruits of subsequent harvests, and care must be taken not to damage the tree, which would make it easy for parasitic fungi to penetrate the tissues of the tree.*

Opening the pods

The pods are opened to remove the seeds. This operation must be completed no longer than six days after harvesting.

In general, the harvested pods are grouped together and split either in or at the edge of the plantation. As this is always done at the same place, it is recommended that a sufficiently large and deep pit is dug into which to throw all the pod debris, as well as rotten pods collected during harvesting. This will limit the spread of fungal diseases such as black pod. The contents of the pit must be regularly treated when phytosanitary control measures are being carried out.

The simplest way of opening the pods is to use a wooden club which, if it strikes the central area of the pod, causes it to split into two halves; it is then relatively easy to remove by hand the seeds which are attached to the central placenta. Although it is a widespread practice, the use of a cutting tool, such as a machete, for example, is not recommended as the seeds can be damaged.

When the harvest is large, pod-opening is a very labour-intensive task. If, for example, one man can gather 1500 pods in a day, it is also one day's work to open them. Therefore, for a long time, attempts have been made to mechanise this operation. Several prototypes of pod-opening machines have been tried, such as the Cacaoette, made in France, the Zumex from Spain or the Pinhalense from Brazil but, because of their purchase price, their running costs and the skill needed to operate them, they are not yet widely used. A manual pod-opening machine, which is lightweight and practical, has been tried with more success by Cameroonian planters.

Preparation of commercial cocoa

In order to be sold as 'cocoa beans', the fresh seeds removed from the pods have to undergo two very important processes – fermentation and drying. The production of high-quality cocoa, which is so much sought after today, depends directly on how these two processes are carried out, the main aims of which are:

- to remove the mucilaginous pulp which surrounds the seeds;
- to cause the death of the embryo and consequently prevent it from germinating;
- to bring about complicated biochemical changes inside the cotyledons, leading to a reduction in the bitterness and astringency and enabling the 'precursors' of the chocolate flavour to develop;

- finally, to reduce the water content of the fermented beans from approximately 60 per cent to 6–7 per cent, in order to block the enzymatic reactions and to enable the commercial product to be stored safely, free from pests and diseases.

Fermentation

Process and duration of fermentation

Placing the fresh seeds in specially-prepared containers, or more simply in piles, rapidly triggers the fermentation of the sweet mucilaginous pulp which surrounds each seed.

This pulp, which has an acid pH due to the presence of citric acid, provides a very suitable environment for yeasts to develop. These convert the sugars into alcohol, giving off CO_2 and producing a sharp rise in temperature (up to 35–40°C after 48 hours). At the same time, the pH rises and the lactic acid bacteria start to develop.

Hydrolysis of the pectins in the pulp is accompanied by the draining-away of the surplus liquid and the aeration of the beans; this allows the acetic bacteria to intervene and develop rapidly. By oxidation, these convert the alcohol into acetic acid. Regular stirring of the entire mass is necessary to promote aeration and to obtain even fermentation; stirring should generally be carried out every 48 hours.

The combined production of acetic acid and the sharp rise in temperature kills the embryo of the seed, the cells of which become permeable. The reserve cell enzymes are therefore enabled to come into contact with the polyphenols of the pigment cells. The anthocyanic pigments are first of all hydrolysed during an initial anaerobic phase, then oxidised in the second aerobic phase. The cotyledons then show the characteristic brown colour of cocoa.

At the same time, almost 40 per cent of the theobromine present in the fresh cotyledons is concentrated, by diffusion, into the teguments of the bean, thus producing a reduction in the bitterness of the 'well-fermented beans'.

The duration of the fermentation process varies widely, depending on:
- the genetic structure of the selection
- the climate;
- the volume of the mass of cocoa undergoing fermentation;
- the method of fermentation adopted.

Fermentation is arrested when a certain number of typical changes occur at approximately the same time: namely, the beans swell, they develop a certain odour, the cotyledons turn brown and the temperature falls. This occurs on average after four to six days for cocoa of the Forastero and Trinitario type and after two to three days for the Criollo type.

> The most important change which occurs in the cotyledons during fermentation is the appearance of the precursors of the chocolate flavour. These substances which, among others, contain free amino acids and monosaccharides, are capable of giving the cocoa beans, after roasting, the characteristic flavour and aroma sought after in this product.

Traditional methods of fermentation

In *baskets*: the baskets can be of any size, and are woven from vegetable fibres; they may contain quantities of seed ranging from 10 to 150 kg. The filled baskets are placed on the ground or on tables and are covered with banana leaves. The beans are mixed by transferring them from one basket to another.

In *heaps*: the seeds are placed on a carpet of banana leaves, which are themselves placed on a bed of branches which help the surplus liquid to drain away. The heap of seeds is covered with banana leaves.

In *boxes*: the volume of the boxes varies enormously and must fit in with the harvesting and plantation management conditions. Small wooden boxes with internal dimensions of 50 × 50 × 50 cm, which are capable of holding 80–85 kg of seeds provide good conditions for fermentation to take place. Four boxes of this type will be needed to treat one tonne of commercial cocoa, i.e. the harvest from approximately one hectare.

Plastic fermentation tanks have been used, particularly in Côte d'Ivoire. These tanks, which can be stacked to facilitate mixing, each contain 80 kg of seeds.

In large plantations, a series of large boxes which can hold more than one tonne of seeds is generally used. The fermentation boxes must have drainage holes and be well-ventilated. When the box is full, the cocoa is covered with banana leaves so as to encourage the natural seeding of yeasts and bacteria. Mixing is carried out by transferring the beans from one box to another. This operation is made easier by pulling back movable partitions and by arranging a series of boxes to be stacked in a cascade formation.

Factors involved in fermentation

- The degree of ripeness of the pods
  Seeds from ripe pods ferment normally, whereas those from unripe pods do not ferment correctly. Overripe pods may contain germinated seeds.
- Pod diseases
  In the event of disease, harvesting should be carried out more frequently. All the pods are harvested, but only the ripe pods whose

beans are not affected are used for the preparation of commercial cocoa. The others are discarded. Unfortunately, this principle is adhered to in far too few producing countries.

- Types of cocoa
  As we have seen, the duration of the fermentation period for the Criollo cocoas is relatively short – two to three days – whereas fermentation lasts from four to six days and sometimes more with the Forasteros and Trinitarios.

- Climatic and seasonal variations
  The relative weights of pulp and sugars per seed vary considerably, according to the growing conditions. For example, it is well-known that seeds will have a limited amount of pulp during the dry season. Elsewhere, too much pulp, which may develop under certain conditions, reduces gaseous exchange which, together with a high sugar content, can result in high acid levels in the cotyledons.

  Finally, there are countries where the variations in temperature are such that it may prove to be necessary to insulate the fermentation boxes. At altitudes higher than 800 m, the relatively low temperatures lead to fermentation slowing down.

- Quantity of cocoa
  A minimum quantity of around 70 to 90 kg of fresh seeds is needed to obtain a satisfactory level of fermentation.

- Duration
  When fermentation is complete, the cotyledons of the beans must be of a uniform brown colour. Inadequately fermented cocoa has beans which are still purple; these will give a bitter and astringent product. On the other hand, an excessively long fermentation period runs the serious risk of producing a cocoa with a very bad taste due to the beginning of putrid fermentation.

Drying

The aim of drying is to reduce the water-content of the fermented beans, which is approximately 60 per cent, to less than 8 per cent, so as to ensure that the cocoa is kept in good condition for storage and transportation.

Drying rates and cocoa quality

If drying is too slow, there is the danger of moulds developing, which is a very serious fault for commercial cocoa. On the other hand, if drying is too rapid, oxidation may be prevented and the acetic acid may be retained in the cotyledons, resulting in excessive acidity.

This acidity is due to the presence of volatile and non-volatile acids, of which the most important are acetic, citric and lactic acids. Most of the acetic acid is eliminated during the fermentation stage, but the non-

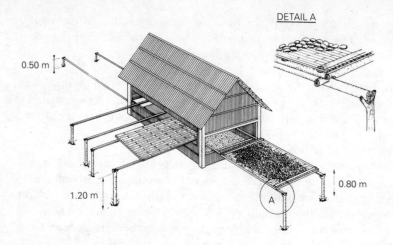

0.50 m

1.20 m

0.80 m

A

Fig 27 *Drying 'autobus' for a family plantation*

volatile lactic acid is retained throughout the drying and processing stages. It is recognised that beans dried in the sun are less acidic than those dried artificially.

During drying, it is essential that, wherever possible, defective beans are removed; this includes flat beans, germinated beans and broken beans, as well as any foreign bodies.

The methods used to dry cocoa can be divided into two main types – natural or sun-drying, and artificial drying.

Natural drying

Sun-drying is the simplest and also the most frequently-used method in most of the producing countries. It does, of course, depend on the climatic conditions and, in general, the beans have to be exposed for one to two weeks.

- Drying on matting

  In small plantations the cocoa is sometimes simply laid out in the sun, spread in a thin layer on mats placed directly on the ground or, more frequently, raised on rough supports to prevent domestic animals interfering with them. After three or four days the cocoa is stirred and dried again. A good method then is to spread it out on a black plastic sheet also stretched out on the mat. The heat absorbed by the plastic then dries the cocoa thoroughly in ten days. Every evening, or as soon as the rain comes, the sheet must be rolled up with the cocoa, which protects the beans and avoids moisture entering the beans.

- The drying 'autobus'

  The drying 'autobus' consists of a shelter, usually made of wood covered with matting or tarpaulins. There are openings on each side which give access to wooden ramps at different heights, on to which drying trays can slide. These trays, which are mats stretched on a wooden framework, can be slid rapidly under the roof. Larger-scale dryers of this type, constructed from more sophisticated materials, are often used in large industrial plantations where several heights of ramp on rails can be placed one on top of each other in order to increase the usable surface area of the trays.

- Movable roof dryer
The principle of the movable roof dryer is the opposite of that of the drying wagon. The drying area, a cement or wooden platform, is fixed. The tarpaulin roof can be moved along rails laid along the length of the drying area.

Artificial drying
When the climatic conditions are not suitable for drying the beans in the sun, or when the size of the plantation is such that large drying areas will be needed at peak periods if natural drying were to be carried out, artificial methods will have to be used.

- Simple dryers
Simple dryers are ovens with a cement floor, or, even better, a tile floor, upon which the cocoa is placed to dry by heating the base of the oven. Hot-air ovens are operated by conducting smoke from an external fire over the beans. With the latter system, the hot air dries the cocoa by permeating through the trays on which the beans are spread.

Among the best-known hot-air dryers, the 'Samoan' dryers, which are relatively simple to make, are used in family plantations as well as in large plantations which use wood or fuel oil for heating.

In all cases, strict precautions have to be taken to avoid the cocoa being contaminated by the smoke. The chimney over the fire must be sufficiently high and sufficiently far away from the roof covering the dryer. Similarly, particular care must be taken to ensure that the drying platform and the smoke ducts are impermeable.

- Mechanical dryers
Many types of mechanical dryer are used on the larger plantations. They are either movable tray dryers circulating in a tunnel through which hot air is blown, or rotary dryers where the hot air passes over the cocoa contained in a moving cylinder. Equipment of this type is only profitable if a large volume of cocoa is to be dried. Drying lasts 10 to 20 hours, depending on the initial moisture-content of the cocoa.

Maize dryers of the 'Scolari' type have also been used successfully. The layer of beans is subjected to a stream of hot air and the cocoa is agitated mechanically in order to dry the beans evenly.

- Automated workshops
  One of the first fully-automated workshops for the fermentation, drying and packing of cocoa has been designed and put into practice by the IRCC in Côte d'Ivoire. This experimental unit can receive 20 to 25 tonnes of fresh seed per day and, at the end of the programme, produces a cocoa with a good flavour and a good aroma.

  In countries where cocoa production is mainly a family business – in Africa, for example – the operation of a workshop of this kind depends fundamentally on the regularity and volume of the supply, which must be planned and adhered to strictly.

Recovery

'Recovery' is the term applied to the ratio between the weight of the dry cocoa obtained after fermentation and drying and the weight of the fresh seeds collected during pod-opening. The recovery rate varies according to the type of cocoa used, the degree of ripeness of the pods, the climatic conditions existing during harvesting, the duration of and conditions under which fermentation took place and the drying conditions. There can, therefore, be extremely wide fluctuations in the recovery rates, ranging from 32 per cent to 46 per cent, but an average figure of 40 per cent is fairly generally used when an estimate of the yield has to be made.

Cleaning and bagging

After drying, the beans are packed in jute bags. This is often preceded by the beans being sorted, and the elimination of any which are flat or broken, any which have germinated and any other impurities, in order to improve the quality of the product. The international standards state that the net weight of a cocoa bag must be 62.5 kg, i.e. sixteen bags to 1 tonne.

Storage of commercial cocoa

Cocoa beans have to be stored by the producer before delivery, the exporter before export and the industrial processor before use. Cocoa is a fragile commodity which, during storage, is exposed to various risks of deterioration: the most usual ones being the impregnation of odours and the development of off-flavours, an uptake of moisture and the development of moulds and insect attacks. International standards stipulate certain storage precautions:

- ambient humidity must not exceed 70 per cent;
- the bags must be stored at least 7 cm from the ground, normally on a duckboard floor which allows air to circulate freely;

- there must be a passage at least 60 cm wide between the walls and the bags and between bags of different types of cocoa;
- if necessary, disinfection by fumigation with methyl bromide and/or by spraying pyrethrin-based insecticides. Protection against rodents and other pests must be provided:
- steps must be taken to avoid contamination by odours, off-flavours or dust;
- periodic checking of the moisture content of each lot must be carried out.

7 Working times per hectare

It is practically impossible to define standard working-times per hectare of plantation which would be of value as international averages.

Situations vary considerably from one country to another or even from one region to another. There is a huge number of factors to be taken into account and they vary enormously between the mosaic of small family plantations (in Africa, for example) and the large industrial plantations of South East Asia.

Among these elements, the planter himself, with his characteristics, his organisation, his skill, the time and labour he has available, as well as his obligations and customs (or habit), are of prime importance. Then there are other elements which are also very variable, such as the topography of the land to be used as the cocoa plantation, the vegetation covering it and its surrounding ecological factors. One may easily understand the difficulties and the many visits to the land which even the smallest socio-economic study on the operation and the benefit of cocoa-growing would involve. The departures from any conclusion aiming to be national, regional or universal in this field are almost as numerous as the planters themselves.

The data shown here come from several sources and are therefore only of value as a general indicator. The planter will be able to draw up his own schedule by adjusting these figures according to his means and working conditions.

Table 5 *Estimate for the working time for 1 hectare of cocoa trees in forest conditions (working days)*

– Density of planting: 1333 plants/hectare (spacing 3 m × 2.50 m)	
– Temporary shading at the same density (plantain bananas, *Gliricidia*)	
– No cover or food crop	
– One working-day = 5 to 6 hours	
– *Nursery* (80 m² for 2000 plants)	
– Constructing the shading	4
– Establishing the beds	2
– Supplying the soil (5 to 6 m³)	6
– Filling the bags (300/day)	7
(bags 12 cm in diameter × 25 cm high)	
– Sowing the seed	2
– Maintenance – watering (6 months)	25
Total time in nursery (days)	46
Preparation of the forest land	
– Boundary-marking	3
– Slashing undergrowth	13 to 20
– Felling – chain-sawing	50
– Marking paths and the drainage or irrigation system	60
(150 m of paths/hectare)	
– Extracting stumps and creating the windrows	50 to 66
– Burning the windrows	10
– Marking out for cocoa trees:	7
cutting the stakes (200/day)	
– staking out	14
– Marking out for temporary shading:	
plantain bananas: cutting the stakes	7
staking out	7
Gliricidia: preparing cuttings	7
planting	14
– Digging planting holes (40 cm³)	27 to 34
(no hole-digging for temporary shading)	
– Filling in the holes	7
– Realigning the stakes after hole-digging	7
Total time spent on land preparation (days)	283 to 313

Table continued on next page

Planting the trees
- Preliminary weeding — 2 to 8
  (manual weeding: 8 days, or chemical weeding: 2 days)
- Transporting the plants — 10
- Planting — 18
- Palm frond shelter construction — 18
- Inspection and replacement (10 per cent) — 8

Planting plantain bananas
- Preliminary weeding — 2 to 8
- Transporting plants — 10
- Planting — 27
- Nematicide spraying — 3
- Inspection and replacement — 5

Planting Gliricidia
- Performed at the same time as staking out — 0

Total planting time (days)	103 to 115

Cultural operations
For each of the first three years:

- Weeding (if a herbicide is used, 6 rounds = 12 days, if — 12 to 42
  cutting back, 7 rounds = 42 days)
- Formation pruning (2nd jorquette) and sucker removal
  (in 6 rounds) — 16
- Removal of unwanted vegetation and creepers — 10
- Insecticidal treatments (in 4 rounds) — 8
- Fertiliser application (in 3 spreadings) — 6
- Harvesting bananas — 16
- Removal of banana shoots (in 3 rounds/year) — 9

Total annual maintenance period for the first three years (days)	77 to 107

For each subsequent year (after obtaining closed, integrated canopy)

- Weeding (in 4 rounds) — 16
- Cutting back banana plants — 4
- Sucker removal (in 8 rounds) — 8
- Insecticidal treatments (2 double rounds if necessary) — 8
- Fungicidal treatments — (as required)
- Fertiliser application (in 3 applications) — 6

Total annual maintenance period after obtaining closed, integrated canopy (days)	42 + fungicidal treatments

Harvesting – Fermenting – Drying
(for 1 tonne of commercial cocoa)

– Harvesting 1500 pods per day	17
– Opening 1500 pods per day	17
– Fermentation and drying	16
Total harvesting, fermenting and drying (days)	50

8 Qualities and defects of commercial cocoa

The chocolate-manufacturer, who is the main user of cocoa beans, requires a product which meets his specific manufacturing requirements. The cocoa must be:

- dry;
- high in fat content;
- uniform, with beans weighing on average at least 1 g. Uniformity of the lots is an important criterion for the chocolate-manufacturer when selecting the beans.

Above all, he wants to buy a product which will enable him to obtain the characteristic flavour of chocolate, i.e. all the flavours and aromatic qualities which, unfortunately, it is very difficult to define objectively and which can only be presumed to exist upon a physical examination of the beans.

Species and soils

These are the main factors used today to classify cocoas as either bulk or fine cocoas.

Bulk cocoas

Bulk cocoas come from Forasteros or hybrids of Forastero trees. They represent 90 to 95 per cent of world production and vary enormously in quality, depending on their origin and the preparation methods (fermentation and drying) used in the different countries.

Fine cocoas

Fine cocoas are produced by Criollo and Trinitario trees. They are found mainly in Indonesia, Papua New Guinea, Samoa, Sri Lanka, Trinidad and in a number of other Caribbean Islands. These are cocoas with a pale shell and the beans are more or less dark cinnamon-coloured after fermentation and drying, with a particularly sought-after flavour.

However, at any stage in its production, a cocoa's characteristics can be changed by external factors which sometimes lead to very serious variations in quality. Table 6 shows the main defects of commercial cocoas in decreasing order of severity.

Table 6 *The main defects of commercial cocoas*

Defects observed	Causes
Mouldy beans (major defect) A mouldy flavour in the finished product with the possible presence of mycotoxins.	– Pods harvested when overripe with germinated seeds. – Seeds damaged during pod-opening. – Excessively long fermentation period. – Inadequate drying. – Renewed moisture uptake during storage.
Slaty beans Very astringent, very bitter chocolate, without the typical chocolate flavour.	– No fermentation.
Purple beans Poor quality chocolate with little flavour, astringent and bitter	– Under-fermentation.
Very dark brown beans with black marks Very unpleasant off-flavours.	– Excessively long fermentation, with the beginning of putrid fermentation.
Beans with a smoky flavour Typical unpleasant flavour.	– Inadequately protected from smoke during drying and/or storage. – Beans stored close to strongly-smelling foods (smoked meat or fish).
Insect-damaged beans	– Beans stored for too long without being regularly checked or given appropriate treatments against insects.

Method used to assess the quality of cocoa beans

Only some characteristics of cocoa beans can be assessed by objective methods. For example, the moisture content can be evaluated and the cocoa butter content can be measured, as can pesticide residues.

However, with the exception of the evaluation of the moisture content, in the market itself a cocoa is assessed only by subjective methods usually limited to what is referred to as the 'cut test', i.e. a visual examination of a cross-section of the cotyledons, sometimes supplemented by a testing.

Samples are taken at random from at least one bag out of three. The stab sampler must be inserted first in the upper part, then in the middle and then near the bottom of the bag. At least 300 beans per tonne or fraction of a tonne of commercial cocoa have to be tested.

> *Cocoa of merchantable quality must have undergone even fermentation and drying. It must have a moisture content of less than 8 per cent. The product must not contain any foreign bodies or live insects, or any bean with a smoky or any other foreign odour, and must not show any signs of deterioration. The beans must be reasonably uniform in size and there should be no broken beans or pieces of shell.*

Cocoa quality standards

Commercial grades are drawn up according to the percentages, which must not be exceeded, of faulty beans found during the cut test. At present, there are four main grading systems in use for commercial cocoa, as shown in Table 7.

Table 7 *Grading systems for commercial cocoa (percentages)*

1 International standards (FAO); applied by most of the producing and consuming countries:

	Mouldy beans	Slaty beans	Other sub-standard beans (insect-damaged, flat or germinated)
Grade I	3	3	3
Grade II	4	8	6
Ungraded	>4	>8	>6

2 Standards adopted by Brazil:

	Mouldy beans	Slaty beans	Insect-damaged-beans	Germinated beans and other defective beans
Superior	2	2	2	2
Bom	4	4	4	4
Abaiso do prado (export subject to special authorisation)	8	8	5	10

3 Standards adopted by the AFCC* and the CAL*:

	Slaty beans	Other defective beans (mouldy, insect-damaged, germinated)
Good fermented	5	5
Fair fermented	10	10
Fair average quality		12

4 Standards adopted by the CMA*:

Mouldy beans	Other defective beans (insects, etc.)	All the defects
4	4	6

*AFCC = Association Française du Commerce des Cacaos – France
 CAL = Cocoa Association of London – Great Britain
 CMA = Cocoa Merchants Association – United States

Other characteristics of commercial cocoa

Purity
The products of the chocolate-manufacturing industry must, like other food products, be pure and uncontaminated. National and international public health authorities are more and more concerned about the purity of foods and their ingredients. In the case of commercial cocoa, the main sources of impurity are pesticide residues, bacteria, several species of insect and foreign bodies.

Uniformity
The manufacturers aim to produce chocolates of unvarying quality. It is natural, therefore, that they should prefer cocoa from sources which are

able to guarantee not only cocoa of constant and as high a quality as possible, but also a regular supply of beans.

Yield in terms of edible matter
The yield of the most useful part of the cocoa bean has a direct effect on its value to the manufacturer and, therefore, on the purchase price which he will be prepared to pay. A certain number of factors which can be measured objectively affect the quantity of edible matter, i.e. the quantity of untreated beans as well as the quantity of the most valuable part, the cocoa butter, which can be extracted from one lot of cocoa.

These factors are:

• the size and uniformity of the beans: one bean must weigh at least 1 g and in one lot not more than 12 per cent of the beans must be outside the range of plus or minus one-third the average weight;
• the fat content: from 50 to 58 per cent;
• the water content: less than 8 per cent;
• the presence of foreign bodies, flat beans or insect-damaged beans.

Summary of the chocolate manufacturer's requirements

In order to produce high-quality chocolate, the manufacturers try to obtain beans with the following characteristics:

• Beans which are likely to develop a marked chocolate aroma after processing.
• Absence of off-flavours, such as mouldy or smoky flavours, or excessive acidity and astringency.
• Beans which can easily be classed as Grade I according to the international (FAO) standards. Cocoa which does not meet this standard is not suitable for manufacturing high-quality chocolate.
• Uniformity of bean size and, on average, weighing at least 1 g per bean.
• Good fermentation and careful drying, with an absolute maximum moisture content of 8 per cent.
• Uniformity.
• Absence of impurities: pesticide residues, bacteria, live insects and foreign bodies.
• A free fatty acid content of less than 1%.
• A fat content of 50–58 per cent, shell content of 11–12 per cent and a hard cocoa butter.

9 Manufacture of chocolate products

Semi-finished products

These are cocoa mass and its direct derivatives, cocoa powder and cocoa butter.

Cocoa mass production

Preliminary processes
When it arrives at the factory, commercial cocoa is first cleaned and sorted by passing it over a continuously vibrating screen, which is very well aerated and is filled with powerful magnets. Any metallic foreign bodies, dusts and broken beans are removed.

Roasting
Roasting is a basic operation which enables the moisture content to be brought down to 1.5–2 per cent. It leads to the evaporation of the volatile acids (acetic acid), assists in separating the shell from the nib and, in particular, promotes the development of the chocolate flavour.

Roasting has to be a very precise procedure, the intensity and duration of which must be carefully regulated according to the particular characteristics of the beans being processed. It is usually a continuous process, using a temperature of 100 to 150°C for 20 to 40 minutes. The beans are then cooled rapidly.

Winnowing
The cooled beans are then transported into a winnowing machine, or cocoa-breaker, where the shells are first cracked and then separated from the nibs by air currents. The nibs and nib fragments, or 'cocoa grains', and the germs are then separated by density on vibrating screens.

Blending and grinding
When it is necessary to have a blend of beans of different origins, the composition of which remains the secret of each chocolate manufacturer, this has to be done before grinding takes place.

The cocoa nibs are finely crushed at a relatively high temperature (50 to 70°C) in cylindrical grinders which move closer and closer together. As the cocoa butter melts, the high temperature enables a fluid mass to be obtained, the fineness of which is one of the conditions determining the quality of the product obtained, which is called cocoa mass.

The cocoa mass can be kept fluid under hot conditions or moulded and cooled to be stored. It is the raw material for conversion into commercial cocoa. It is often made in the producing countries and exported in this form. It will be used to make cocoa butter, powder or chocolate.

Making cocoa butter and powder

Depending on the raw material used to make the butter (whole beans, cocoa nib or cocoa mass), and depending on the extraction process used, distinctions can be drawn between:
- Pressed cocoa butter or cocoa butter: the liquid mass is pressed in hydraulic presses at pressures up to 600 kg/cm^2
- Extruded cocoa butter: the cocoa butter is obtained by extrusion, the hydraulic presses then being replaced by an expeller screw.
- Refined cocoa butter: this is obtained by pressure, by extrusion using expeller screw, by extraction using a solvent or by a combination of these processes, and is then refined.

The cocoa butter from the press is filtered and, if necessary, is neutralised and refined, deodorised and tempered, i.e. kept for a time at a temperature close to its melting point, to allow for the initial stable formation of the crystals. It is then moulded and cooled. At this stage it is hard in consistency, waxy, slightly shiny, pale yellow in colour and oily to the touch. It melts at a temperature close to 35°C, giving a clear liquid.

Cocoa butter is a mixture of glycerides of oleic (37 per cent), stearic (34 per cent), palmitic (26 per cent) and linoleic (2 per cent) acids.

The cake, which is the residue left at the bottom of the presses after the cocoa butter has been extracted, contains a further 20 per cent butter. It will be converted into cocoa powder by winnowing and grinding. Further processing will give a so-called 'low-fat' or 'defatted' cocoa powder, containing only 10 per cent of cocoa butter.

The finished products: chocolates

Chocolate powder

Chocolate powder is made directly from cocoa powder to which sugar (68 per cent maximum) and usually vanilla are added.

Soluble cocoa

Cocoa powder is not normally soluble in water and will float on the surface. The addition of alkaline salts will produce a suspension, but a more homogeneous mixture which will disperse in liquids such as milk or water can be produced by adding lecithin, or by an agglomeration of the powder particles in order to make them more dense. This preparation is referred to as soluble cocoa.

'Breakfast chocolate'

'Breakfast chocolate' is characterised by the presence of greater or lesser additions of precooked meal to the cocoa powder and to the sugar.

Chocolate

Chocolate is a mixture of cocoa mass and sugar to which cocoa butter, milk, fruit or flavourings can be added.

A mixture of mass and sugar: cocoa mass, which is kept fluid by heat, and sugar, very finely ground beforehand, are mixed in a blender under vacuum. The proportions are carefully programmed and the mixing is carried out automatically.

Refining: refining gives an absolutely homogeneous mixture and a very fine grain size. It is carried out in cylindrical grinders which are placed one on top of the other and which are adjusted to operate at increasingly closer spacings, rotating at differential speeds of around 200 revs/minute. The mass then becomes dry and flaky. It is kneaded again in a blender and it is at this stage that the cocoa butter which is required to impart the finest qualities, as well as the vanilla or the different aromas necessary to give a specific flavour to the chocolate are added.

Conching: conching is one of the most important operations in making chocolate and, to a great extent, the quality of the product depends on it, both in terms of flavour and aroma. It is carried out in large vats – the conches – where the mass is continually mixed and kneaded for 24 to 72 hours at a temperature varying between 60 and 80°C. The time spent in the conches influences the production of the velvety texture and smoothness of the chocolate. Most of the cocoa butter and lecithin needed is added towards the end of conching.

Tempering: tempering consists of bringing the mixture down to a temperature of between 28 and 31°C in automatic tempering vats so as to obtain stable crystallisation of the cocoa butter.

Dressing: dressing includes moulding, where the tempered chocolate passes into a weighing hopper which distributes it into moulds, tapping, which causes the moulds to be continually shaken in order to distribute the mass evenly and remove air bubbles, refrigeration in refrigerated tunnels at approximately 7°C and, finally, removing the chocolate from the moulds, which is done by turning out the moulds on to a felt conveyor belt which receives the bars.

Packing and packaging: these final stages are also fully automated.

Some legal definitions

Sweetened powdered cocoa, sweetened cocoa and chocolate powder
A mixture of cocoa in powder form and sugar, containing at least 32 g of cocoa powder per 100 g.

Cooking or household chocolate
Contains at least 30 g of total dry solids of cocoa per 100 g, 18 g of which is cocoa butter, the remainder being sugar.

'Milk' chocolate
The milk in powdered or concentrated form (milk containing at least 24 per cent fat) is incorporated before refining. 100 g of milk chocolate must contain at least 25 g total dry solids of cocoa, of which 2.5 g is defatted dry cocoa, 14 g at least of solids deriving from the evaporation of milk, at least 3.5 g cream and a maximum of 55 g sugar.

'Enrobing' chocolate
This must contain at least 31 per cent cocoa butter. It is used in confectionery and biscuit-making.

White chocolate
This is made of milk, cocoa butter and sugar.

'Chocolate' or 'cocoa' products
They must contain at least 35 per cent mass or cocoa powder if they are solid, 32 per cent if they are in powder form and 6 per cent if they are liquid.

Chocolate-flavoured or cocoa-flavoured products
Contain 20 to 35 per cent cocoa mass or cocoa powder.

Products with a 'cocoa flavour or aroma' or with a 'chocolate flavour or aroma'
Contain 0 to 20 per cent cocoa mass or cocoa powder.

There are four categories of bar chocolate: plain chocolate, milk chocolate, white chocolate and filled chocolate.

The categories plain chocolate and milk chocolate are themselves subdivided into four categories called: cooking chocolate, eating chocolate, enrobing chocolate and so-called luxury chocolate, according to the minimum content of cocoa components.

The higher the cocoa content of a chocolate, the lower its sugar content, with almost always an increase in the percentage of cocoa butter. The higher the cocoa content the more bitter the chocolate. The more sugar the chocolate contains, the less bitter it is. The more cocoa butter it contains, the more oily it is, and the less it contains, the more brittle it is. The consumer will therefore be able to ascertain the nature of the product on offer by the description of the bar and the compulsory reference to the cocoa content.

We should mention here the particularly precise and strict nature of French legislation on the definition, of chocolate, which, it states, is 'the product obtained from cocoa grains, in mass form, in powder form or low-fat cocoa in powder form and sugar, with or without the addition of cocoa butter'.

Hazelnuts, almonds or walnuts can be added to the chocolate as can milk, flavourings or natural aromatic substances and ethyl vanillin, provided that the composition of the finished product is stated and that the legal definitions have been used.

At this time of change in Europe, when several countries have modified this 'noble' composition of chocolate by replacing the cocoa butter with other animal or vegetable fats, it would appear to be important, in the interest of the consuming countries and particularly that of the producing countries, to adhere closely to these French legal provisions.

Table 8 *The minimum cocoa content of chocolate (%)*

Plain chocolates	
Cooking chocolate	30
Plain chocolate	35
Dark enrobing chocolate	16
Luxury/extra-fine-chocolate	43
Milk chocolates	
Milk cooking chocolate	20
Milk chocolate	25
Milk enrobing chocolate	25
Luxury/extra fine milk chocolate	30

10 Nutritional value

Pleasant to eat and drink, the products of the chocolate-making industry have a high energy value in relation to their volume. They also contain useful minerals, vitamins and stimulating substances. Chocolate today has a complex composition; some 800 components, at least, have so far been counted. Table 9 gives the average composition of a 100 g bar of chocolate, showing its main components. Chocolate is therefore a complete food, particularly high in energy due to its high levels of glucides, lipids and protides. A 100 g bar has 500 calories. It contains theobromine and caffeine, which are alkaloids capable of stimulating the central nervous system. Chocolate also contains trace elements, is rich in vitamins, and is also a source of mineral salts. Its high magnesium and iron content may, if necessary, reduce any deficiency of these elements in the diet. Furthermore, chocolate also contains a substantial amount of phosphorus.

The properties of chocolate can, essentially, be summarised as having restorative, energy-producing and tonic effects on the body. Many of chocolate's components form part of these functions, particularly the theobromine. Its stimulating effect on the central nervous system is so great that it can be said to have a 'doping effect' (theobromine is prohibited by law in competitive sports).

Finally, a very recent study carried out at the University of Compiègne shows that the plain chocolate eaten in France has an average cholesterol content of 1 mg per 100 g (10 mg/100 g for milk chocolate) and therefore plays only a negligible role in cholesterol intake. Chocolate should therefore no longer be banished from diets connected with the risk of atherosclerosis.

The contra-indications which are associated with chocolate lie in its high sugar and fat content. Diabetics, the obese and, by extension, hyperuricaemics and gout sufferers are advised not to eat it. Chocolate, which is always rich in oxalic acid, should also not be eaten by people suffering from oxalic lithiasis.

The consumption of chocolate in reasonable quantities, i.e. less than 50 g per day, leads neither to biliousness nor to migraine.

Table 9 *Average composition of a 100 g bar of chocolate*

Components	Plain chocolate	Milk chocolate
Nutritional components		
Proteins	3.2 g	7.6 g
Lipids	33.5 g	32.3 g
Carbohydrate	60.3 g	53.0 g
Pure lethicin	0.3 g	0.3 g
Alkaloids (particularly theobromine)	0.6 g	0.2 g
Minerals		
Calcium	20 mg	220 mg
Magnesium	80 mg	50 mg
Phosphate(s)	130 mg	210 mg
Trace elements		
Iron	2.0 mg	0.8 mg
Copper	0.7 mg	0.4 mg
Vitamins		
A	40 IU	300 IU
B_1	0.06 mg	0.1 mg
B_2	0.06 mg	0.3 mg
C	1.14 mg	70 IU
D	50 IU	70 IU
E	2.4 mg	1.2 mg
Assimilable energy		
Calories	495	515

Glossary

Adventitious plant A plant which grows on land under cultivation without having been sown (casual weed).

Androecium All the male organs, stamens and staminodes of one flower.

Anticryptogamic (= fungicide) Chemical product used to kill fungi.

Bean The cocoa bean is a seed which has undergone fermentation and drying.

Budding Opening of the bud and the first appearance of the leaves.

Cherelle Young fruit of the cocoa tree during its development.

Chromosome Body visible in the nucleus of the cell at the time of cell division. Each chromosome is present in pairs and carries the genetic code of numerous characteristics. The cocoa tree has 2×10 different chromosomes, so the diploid number is $2n = 20$.

Clone Propagated by vegetative or asexual techniques.

Collar Point at which the stem and the root join.

Cortex (or pericarp) Outer tissue of the fruit made up of three separate layers: from the outer to the inner layer, this comprises: the epicarp, the mesocarp and the endocarp.

Cotyledon First formed leaf, usually in the form of a thickened lobe which is attached to the axis of the plantlet and which contains a reserve of nutrient elements. The seed of the cocoa plant has two cotyledons.

Cultivar Selected variety widely used commercially and normally resulting from controlled hybridisation.

Cutin Waxy substance which comprises most of the external membrane of the epidermal cells.

Deficiency Insufficiency of one or more nutrient elements necessary for the optimum development of a plant.

Diploid A term describing a cell with 2n chromosomes.

Epigeal Type of germination in which, as the hypocotyl axis lengthens, it pushes the cotyledons above the surface of the soil.

Family Group of plants which are related by common characteristics and which systematically constitute the superior unit of the genus.

Flower initiation   Initiation of the development of the beginnings of the flowers.

'Flush'   Period when a shoot is developing with the appearance of new leaves.

Gene   Hereditary factor in the chromosome.

Genotype   Individual of characteristic genetic origin.

Genus   Group of species which have certain common characteristics.

Hardening-off   Gradual removal of shading from a nursery to accustom the plants to light.

Hermaphrodite   A flower with both stamens and a pistil.

Heterozygote   Individual of unstable descent, not of pure stock, giving rise to divergences in genetic characteristics.

Hybrid   Plant resulting from a cross between two individuals differing by one or more fixed characteristics. The hybrids of one and the same cross constitute a 'hybrid family'.

Jorquette   The five framework branches which appear at the top of the trunk of a cocoa tree.

Lignification   Maturing of the wood.

Marking out   Inserting and aligning 1 to 2 m high wooden stakes at the exact spot where the future shade trees and cocoa trees will be planted.

Mesoamerica   Growing area including the southern and eastern parts of Mexico, Guatemala, El Salvador and the western parts of Honduras, Nicaragua and Costa Rica. In the pre-Columbian era, these regions shared a number of cultural characteristics which distinguished them from their neighbours in the south and north.

Orthotropic   A shoot which grows vertically.

pH   Number indicating the acidity of soil or a liquid: acid below pH7, alkaline or basic above, neutral at pH7.

Photosynthesis   Property which chlorophyll-containing plants have of breaking-down carbon dioxide and, with the aid of its constituents, carrying out organic syntheses using light energy, resulting in the production of carbohydrates and the release of oxygen.

Phyllotaxy   Arrangement of the leaves on the stem according to a leaf spiral. A 3/8 phyllotaxy means that three revolutions around this spiral have to be followed before two leaves are found in an identical position on the stem. During these three revolutions, eight leaves are included in the spiral.

Pinnate   The petiole of a leaf which continues into a central vein from which secondary veins regularly lead, arranged like the barbs of a feather.

Plagiotropic   A shoot which grows horizontally.

Pod   Mature fruit of the cocoa tree.

Self-incompatibility   Situation where it is impossible for the flowers of a plant or group of plants to be fertilised by their own pollen.

Staminode   Organ of the same origin as the stamens but without the pollen sac and very different in shape.

Stoma   Modified epidermal cell used for gaseous exchanges and water vapour emission.

Superior   Ovary situated above the apparent plane of insertion of the floral parts.

Trace element   Element which has either a favourable or unfavourable effect at very low concentrations, such as boron (B), iron (Fe), copper (Cu), zinc (Zn) or manganese (Mn).

Wind-row creation   Piling up in rows the vegetable debris created during land clearance.

Further reading

Books

Are, L.A., Gwynne-Jones, D.R.G., *Cacao in West Africa* (Oxford University Press, London, 1973).

Braudeau, J., *Le cacaoyer* (G.P. Maisonneuve et Larose, Paris, 1969).

Lavabre, E.M., *Insectes nuisibles des cultures tropicales* (G.P. Maisonneuve et Larose, Paris, France, 1970).

Loué, A., *Etudes des carences et des déficiences minérales sur le cacaoyer* (IRCC/ CIRAD, Paris, 1961).

Wood, G.A.R., Lass, R.A., *Cocoa* (Longman, London, 1985).

Periodicals

Café – Cacao – Thé, IRCC/CIRAD, Paris.

Cocoa Grower's Bulletin, Cadbury Schweppes plc, Bournville, Birmingham.

Conférences internationales sur la recherche cacaoyère, Alliance des Pays Producteurs de Cacao (Alliance of Cocoa Producing Countries), Stephen Austin and Sons Ltd, Hertford.

Marchés tropicaux et méditerranéens, R. Moreux et Cie, Paris.

Index

acid levels
   fermentation 71, 73
   drying 73–4
   in soil 25
Acosta, José de 3
Africa, cocoa production 3, 4, 6
Albizia trees 41
aleurone 20
Almeida cocoa 9
Amelonado cocoa 8
   flowering 13
   planting 43
   self-compatible 16
Amelonado pods 18
androecium 15, 95
angoletta pods 18
anthocyanin 20
anthracnose 55
ants, leaf-cutting 68
areca palms 41
Arriba cocoa 9
Asia–Oceana, cocoa production 6
atmospheric humidity 23
automated workshops 76
Aztec Indians 1

bagging 76
bananas, for shade 39, 42
bases, exchangeable 26
beans 95
   cleaning 76
   defective 74, 76, 83, 84, 85
   quality 82–3, 84
   recovery 76
   storage 76–7
beetles 65
black pod 52, 70
blending 87–8
borers 54, 55, 65, 67

boron deficiency 49
branches
   damaged 66–7
   fan 11–12
   orthotropic 11, 12, 13
   plagiotropic 11, 13
   secondary 12
   see also jorquette
breakfast chocolate 89
brown beans, marked 83
budwood 32
bulk cocoas 82
by-products 6

caffeine 21, 92
calabacillo pods 18
calcium 26
Calonectria 63
calories 93
canopy shading 40, 42, 47
Caribbean cocoa production 6
cassava 39
castor-oil plant 39
caterpillars
   defoliating 36
   wood-boring 67
Catongo cocoa 9
Central America, cocoa 1–3
Centrosema pubescens 40
ceratocystis wilt 65
charcoal pod rot 53
cherelle 17, 18, 95
chocolate 89–91
   nutritional value 92–3
chocolate bars 6, 90–1, 93
chocolate flavour 70, 72
chocolate powder 6, 88
chocolate products, manufacture 82, 87–91

cholesterol content 92
climatic factors 22–4
clonal selection 27–8
cockchafer 65
cocoa 1–3
    as currency 1, 2
    classifications 7–9
    selection 27–8
    soluble 89
    uses 4, 6
    world consumption 4
    world production 6
cocoa beans *see* beans
cocoa butter
    in beans 84
    in cotyledons 20
    types 88
    uses 4
coca content, chocolate 91
cocoa mass 87–8
cocoa plantations
    choosing site 37
    fertilisers 47–9
    maintenance 46–7
    planting 43–6
    preparing land 37–9
    protection 49
    regenerating 41–2
    shading 39–42
cocoa powder 88
cocoa quality standards 84
Cocoa Swollen Shoot Virus 62
cocoa tree
    aerial system 11–13
    collections 27
    diseases 61–5
    flowering 13–15
    fruiting 15–19
    root system 10
    seeds 19–22
coconut palm 23, 41
cola trees 40
commercial cocoa
    defects 83, 84–5
    preparing 70–7
    qualities 82–5
    storage 76–7
compost 44
Comun cocoa 8–9
conching 89
cooking chocolate 90
cortex 17, 95

Côte d'Ivoire
    automated workshops 76
    fermentation 72
cotyledons 19, 95
    structure 20–1
    white 9
Criollo cocoa 7–8
    cocoa collection 27
    fermentation 71
    fine cocoas 82–3
    pod colour 18
cross breeding 29–30
Crotalaria spp 40
cundeamor pods 18
cuttings 28

damping off 36
defoliation
    beetles 65
    caterpillars 36
depulping 31, 34, 70
dressing 89
drying
    artificial 75–6
    labour time 81
    natural 74–5
    and quality 73–4
drying autobus 74–5

ecology, for cocoa growing 22–6
enrobing chocolate 90, 91
enzyme analysis 28
epigeal germination 21, 95
Erythrinae 39, 41

fan branches 11–12
fat content 91, 92
fermentation 70–3
    climatic variations 73
    labour time 81
    traditional 72
fertilisation 16, 18
fertilisers, mineral 47–8
fine cocoas 82–3
finished products, 6, 88–91
flavourings, for chocolate 91
Flemingia macrophylla 40
flower cushions 14, 69
flowering 13–15
flushes 12, 96
Forastero cocoa 8–9
    bulk cocoas 82

cocoa collection 27
fermentation 71
see also Lower Amazonian
   Forasteros; Upper Amazonian
   Forasteros
fruit 16–19
fruiting 15–16
fumigation 77

gene banks 28
genetic characteristics 28
germination 21–2
Gliricidia spp 34, 39, 41
grafting 28, 32
   for regeneration 42
grinding 87–8
ground cover plants 40

harvesting 69
   labour time 81
herbicides 46–7
   at planting 45
   pre-emergence 35
heterozygosity 29, 96
Hevea oil palms 41
holes, for planting 44–5
host trees 53, 56
husk 19
hybrid selection 28–30
hybrid vigour 29
hybridisation 28–30, 96

in vitro techniques 28
incompatibility 16
inflorescences 13
insect hosts 41
insect-damaged beans 83, 84, 85
insects
   pollinators 15–16
   pests 36, 49–51
interclonal hybrids 28
intercropping 39–40, 41
International Board for Plant
   Genetic Resources 31
International Cocoa Organisation
   4
irrigation 23

jorquette 12, 13, 24, 68, 80, 96

Lagarto cocoa 8
land clearance 33–4

land preparation 37–9
   labour time 79
lateral roots 25, 47
Latin America, cocoa publication 6
leaf 13, 32, 68
leaf-cutting ants 68
Leucaena leucocephala 41
Lower Amazonian Forasteros 8–9
   planting 43
   self-compatible 16

magnesium 26, 48
maintenance 46–7, 80
maize-dryers 75
majuscules 59, 60, 67, 68
Malaysia, high-density planting 43
Maranhao cocoa 8
marking out 44–5, 96
Maya Indians 1, 7
mealy pod rot 54
mealybug 56
melted cocoa 4
micropropagation 28
milk chocolate 90, 91, 93
minerals
   in chocolate 92, 93
   deficiencies 48–9, 93
   in fertilisers 47–8
mirids 53, 66
moisture content
   beans 84
   checking 77
Monilophthora pod rot 58
mouldy beans 83, 84, 85
mulch 45
   natural 47

Nacional cocoa 9
National Cocoa Collection 27
nibs 87, 88
nitrogen 26, 48
nursery 32–6
   labour time 79
nutmeg 41
nutrient elements 25, 26
nutritional value, chocolate 92–3

odours 76, 77
off-flavours 76, 77
oil palm, shading 34
organic matter 25–6
orthotropic branch cuttings 28, 96

ovules 18
oxalic acid 92

papaw 39
Para cocoa 8–9
pectins, in fermentation 71
Pentagona cocoa 8
petioles 11, 13
pH 25, 96
phosphorus 26
photosynthesis 13, 23–4
Phytophthora 36
Phytophthora pod rot 52
pigeon pea 40
plagiotropic branch cuttings 28, 96
plain chocolate 91, 93
plant protection 36, 49–51
plantations *see* cocoa plantations
planting
    density and pattern 43
    labour time 80
plumule 19, 22
pod borer 54
pod-boring caterpillar 55
pod debris 70
pod-feeding moth 56
pod-opening machines 70
pod rots 52, 53, 54, 58
pods 17, 69, 96
    appearance 18–19
    damaged 52–6
    deformed 58–60
    diseases 52, 53, 54, 58, 72–3
    germination capacity 22
    harvesting 69
    opening 70
    ripeness 69, 72
pollination 15–16, 17, 30
polyphenols 20
Porcelana cocoa 8
potassium 26, 48
powdered cocoa 4
propagation, vegetative 27–30
pruning
    canopy opening 47
    diseased parts 59
    drooping branches 58
    roots 46
    suckers 12, 47
psyllids 36, 66
purines 20, 21
purple beans 83

quarantine 31

radicle 19
rainfall 23, 37
Real cocoa 8
recovery 76
refining 89
replanting 42
roasting 87
rodents 36, 60, 69
root pruning 46
root rot 37, 61
roots
    fibrous 26
    lateral 47
    taproot 45
rots
    charcoal pod 53
    mealy pod 54
    Monilophthora pod 58
    Phytophthora pod 52
    root 37, 61
    wet 58

Samoan dryers 75
seed gardens 30
seedlings 32–6
    planting 45–6
    raising 36
seeds
    defective 34
    germination 21–2
    morphology 19
    per pod 17
    preparation 34
selection
    clonal 27–8
    hybrids 28–30
self-incompatibility 16, 97
semi-finished products 4,
    87–8
shading 23–4
    adjustments 46
    and fan branches 24
    nursery 33–4
    palm fronds 45
    permanent 40–2
    temporary 39–40
silver skin 19
slaty beans 83, 84, 85
smoky-flavoured beans 83
snails 36

soil
  chemical properties 25
  nutrient elements 26
  organic matter 25–6
  pH 25
  physical properties 24–5
  types, for cocoa 41
soil maintence 46–7
soluble cocoa 89
South America, cocoa 3
sowing 34–5
Spaniards 2, 3
staking out 39
starch 20
stem-borers 67
stomata 13, 97
stumps 61
  extraction 38–9
  regeneration 42
subsoiling 44–5
suckers 12, 47
sugar content 91, 92
swollen ovary 16, 17
swollen shoot 31, 62

tannins 20, 21
taproot 45
taro 39
temperature 23
tempering 89
Theobroma cocoa L *see* cocoa
theobromine 20, 21, 92
thrips 68
trace elements 92, 93, 97

tracheomycosis 63
tree-felling 38
Trinitarios
  fermentation 71
  fine cocoas 82–3
  pod colour 18
  self-incompatible 16
trunk 11–12

Upper Amazonian Forasteros 9
  self-incompatible 16

vascular streak die-back 64
vegetative material, inter-state
  transfer 31–2
vegetative propagation 27–30
verticillium 63
vitamins 93

warty pod disease 60
water, deficiency 23
wet rot 58
white chocolate 90
wilting 18
wind-rows 39, 97
windbreaks 23
winnowing 87
witches' broom 31, 59
wood-boring caterpillars 67
working times per hectare 78–81

xocolatl 1, 3

zinc deficiency 49